# Test Yourself

# Finite Math

**Karen Zak Benbury, Ph.D.**
Department of Natural Sciences and Mathematics
Bowie State University
Bowie, MD

*Contributing Editors*

**Lynn Yarbrough, M.S.**
Chief Scientist
Mathematical and Computational Sciences
Salinas, CA

**Tony Julianelle, Ph.D.**
Department of Mathematics
University of Vermont
Burlington, VT

**Kathleen H. Poole, M.A.**
Chair, Mathematics Department
Hawken School
Gates Mills, OH

**NTC LearningWorks**
a division of NTC Publishing Group
Lincolnwood, Illinois

**Library of Congress Cataloging-in-Publication Data**
is available from the Library of Congress

A *Test Yourself Books, Inc.* Project

Published by NTC Publishing Group
© 1996 NTC Publishing Group, 4255 West Touhy Avenue
Lincolnwood (Chicago), Illinois 60646-1975 U.S.A.
Manufactured in the United States of America.

6 7 8 9 ML 0 9 8 7 6 5 4 3 2 1

# Contents

# Preface

The purpose of this book is to help you pass your tests in finite mathematics. I have tried to cover those types of questions that are likely to be asked on a test, so that there are no surprises. You can use the questions in two ways: to pinpoint the areas in which you are weak, and to help you recognize the various types of problems. Both skills are important. Also, if you find an area that needs work, practice it until you can do those problems exactly right, and without help. May you have great success!

I would like to thank Fred Grayson for his patience; and the reviewers, Lynn Yarbrough, Tony Julianelle, and Kathleen Poole, for their assistance. We have tried to make this as error-free as possible.

This book is dedicated to my husband, Peter J. Benbury, and our friend Ian.

Karen Zak Benbury, Ph.D.

# How to Use this Book

This "Test Yourself" book is part of a unique series designed to help you improve your test scores on almost any type of examination you will face. Too often, you will study for a test—quiz, midterm, or final—and come away with a score that is lower than anticipated. Why? Because there is no way for you to really know how much you understand a topic until you've taken a test. The purpose of the test, after all, is to test your complete understanding of the material.

The "Test Yourself" series offers you a way to improve your scores and to actually test your knowledge at the time you use this book. Consider each chapter a diagnostic pretest in a specific topic. Answer the questions, check your answers, and then give yourself a grade. Then, and only then, will you know where your strengths and, more importantly, weaknesses are. Once these areas are identified, you can strategically focus your study on those topics that need additional work.

Each book in this series presents a specific subject in an organized manner, and although each "Test Yourself" chapter may not correspond exactly to the same chapter in your textbook, you should have little difficulty in locating the specific topic you are studying. Written by educators in the field, each book is designed to correspond, as much as possible, to the leading textbooks. This means that you can feel confident in using this book, and that regardless of your textbook, professor, or school, you will be much better prepared for anything you will encounter on your test.

*Each chapter has four parts:*

 **Brief Yourself.** All chapters contain a brief overview of the topic that is intended to give you a more thorough understanding of the material with which you need to be familiar. Sometimes this information is presented at the beginning of the chapter, and sometimes it flows throughout the chapter, to review your understanding of various *units* within the chapter.

 **Test Yourself.** Each chapter covers a specific topic corresponding to one that you will find in your textbook. Answer the questions, either on a separate page or directly in the book, if there is room.

 **Check Yourself.** Check your answers. Every question is fully answered and explained. These answers will be the key to your increased understanding. If you answered the question incorrectly, read the explanations to *learn* and *understand* the material. You will note that at the end of every answer you will be referred to a specific subtopic within that chapter, so you can focus your studying and prepare more efficiently.

 **Grade Yourself.** At the end of each chapter is a self-diagnostic key. By indicating on this form the numbers of those questions you answered incorrectly, you will have a clear picture of your weak areas.

There are no secrets to test success. Only good preparation can guarantee higher grades. By utilizing this "Test Yourself" book, you will have a better chance of improving your scores and understanding the subject more fully.

# Functions and Lines

---

## Brief Yourself

In the plane, a fixed horizontal line, called the x-axis, and a fixed vertical line, called the y-axis, are designated. Their intersection is called the origin. A point is represented by a pair of real numbers (x, y), where x denotes the distance and direction of the point parallel to the x-axis, and y its distance and direction parallel to the y-axis. This system is called the Cartesian coordinate system for the plane. The two axes divide the plane into four sections, called quadrants. These are numbered 1 through 4 in counterclockwise order starting with the upper right quadrant, the one in which both co-ordinates are positive.

Three-dimensional space is coordinatized analogously, using three axes, which divide space into eight octants. These axes are labeled x, y, and z. It is important to use a right-handed coordinate system: the forefinger, middle finger, and thumb of the right hand line up with the positive x-, y-, and z-axes, respectively. Points have three coordinates; the first two locate a point in the xy plane, and the third gives the number of units above (or below) this point to the point sought.

There are several forms of the equation of a line: the standard form, $Ax + By + C = 0$; the slope-intercept form, $y = mx + b$, for non-vertical lines; the point-slope form, $y - y_1 = m(x - x_1)$, for non-vertical lines; and $x = a$, for vertical lines. The slope of the line can be found using any two distinct points $P(x_1, y_1)$ and $Q(x_2, y_2)$ on the line using the formula $slope = m = \dfrac{y_2 - y_1}{x_2 - x_1}$. Vertical lines have undefined slope. The x- and y-intercepts of a line are the points at which the line intersects the x- and y-axes, respectively, if they exist. Parallel lines have the same slope. Perpendicular lines have slopes whose product is −1, or one line is horizontal and one vertical.

The notion of a function is basic in mathematics. A function is a rule whereby each element of one set (the domain) is paired with exactly one element of a second set (which contains the range). Besides linear functions, polynomial and exponential functions are common. The notation f(x) is used for the value of the function f at x.

# Test Yourself

1. Plot each of the following points.

   a. (2, –4)

   b. (–1, 2)

   c. (3, 4)

   d. (–2, –3)

   e. (0, 3)

   f. (–4, 0)

2. In which quadrant of the plane does each of the points of problem 1(a) – (d) lie?

In problems 3–6, plot some points, and then sketch the graph in the Cartesian plane.

3. $3x - 4y = 12$

4. $y = -\dfrac{3}{2}x - 4$

5. $y = x - 4$

6. $y = \sqrt{x + 1}$

7. Sketch on the same set of axes.
$$x + y = 4, \quad 2x - y = 5$$

8. The formula giving temperature in degrees Fahrenheit, F, in terms of degrees Celsius, C, is $F = 1.8C + 32$.

   a. What is the temperature in degrees Fahrenheit when C = 30 degrees?

   b. What is the temperature in degrees Celsius when F = 50 degrees?

9. Sketch the graph of the line $2x + 5y = 10$.

10. Find the slope of the line between the two points (–2, 3) and (–4, –5).

11. Find the slope of the line through the points (3, –2) and (3, 4).

12. Find the slope of the line through the points (4, 2) and (–5, 2).

13. Find the slope of the line whose graph is given below.

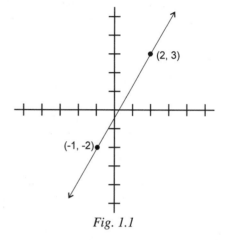

*Fig. 1.1*

In problems 14–21, find an equation of the line from the given information.

14. The line with slope –2 through the point (0, 3).

15. The line with slope 5 through the point (–1, 3).

16. The line through the points (2, –7) and (–1, –1).

17. The line through the points (2, 2) and (–7, 2).

18. The line through (3, –6) and parallel to $y = 4x - 7$.

19. The line through (–2, 5) and perpendicular to $y = -3x + 2$.

20. The line whose graph is

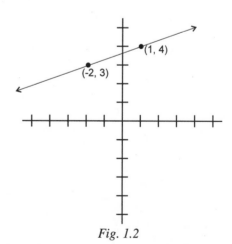

*Fig. 1.2*

21. The line through (2, 3) and perpendicular to y = 17.

22. Find the slope of the line 6y – 2x = 12.

23. Find an equation of the line through (5, –2) and parallel to the line 2x – 3y = 6.

24. Find an equation of the line through (4, –6) and perpendicular to the line 3x + 5y = 15.

25. The cost in dollars of making x widgets is C = 32x + 2000 per month.

    a. How much will it cost to make 800 widgets this month?

    b. The cost of making widgets last month was $26,000. How many widgets were made?

26. f(x) = 2x + 10. Find

    a. f(0)

    b. f(–3)

    c. f(4)

    d. f(a)

    e. f(a + h)

    f. f(2b – 1)

27. $g(t) = \dfrac{t(t-1)}{t+1}$. Find

    a. g(0)

    b. g(1)

    c. g(2)

    d. g(2c)

    e. g(a + 1)

    f. g(b)

28. The values of F(x) are given by the following table.

    | x | 0 | 1 | 2 | 3 | 4 | 5 | 6 |
    |------|---|---|---|----|----|----|----|
    | F(x) | 0 | 4 | 8 | 12 | 16 | 20 | 24 |

    a. Find F(4).

    b. Find F(6).

    c. Find a formula for F(x)

29. A plumber charges $50.00 for a house call and $40.00 per hour. Make a chart showing the amount the plumber charges for 1, 2, 3, 4, and 5 hours of work at a person's house.

30. Find a formula for the amount per house call the plumber in problem 29 charges in terms of the number of hours.

31. To manufacture souvenirs, it costs $1,200 per month plus $2 per souvenir. Find a formula for the monthly cost of manufacturing x souvenirs.

32. Souvenirs in problem 31 sell for $5 each. Find the revenue function giving the amount received if x souvenirs are sold.

33. Find the profit function if x souvenirs are made and sold in a month (problems 31 and 32.)

34. What is the break-even point for the souvenirs of problems 31–33?

35. A car-rental service advertises that it charges $25 a day plus 5 cents a mile. Find the cost of renting a car per day in terms of the number of miles driven.

36. Video Club A charges $2.00 per day to rent videos. Find a formula for C, the cost of renting videos, in terms of the number, d, of the customer's video rental days per year.

37. Video Club B charges a $10.00 membership fee, and $1 per day to rent videos. Find the yearly cost, V, in terms of the customer's videos rental days per year.

38. A $5,000 copier has a useful life of 5 years, and a scrap value of $500. Let t be the age of the copier in years. Find a function, V, giving the value of the copier in terms of its age, t. Assume straight line depreciation.

39. What is the copier in problem 38 worth after

    a. 2 years?

    b. 3 years?

    c. 4 years?

40. What is the annual (straight line) depreciation for the copier in problems 38 and 39?

# Check Yourself

1. **(Cartesian coordinates)**

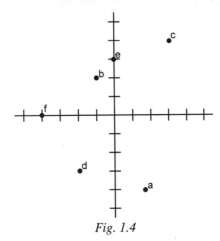

*Fig. 1.4*

2. a. Quadrant IV

   b. Quadrant II

   c. Quadrant I

   d. Quadrant III **(Cartesian coordinates)**

3. $3x - 4y = 12$ **(Cartesian coordinates)**

| x | 0 | 2 | −2 | 4 |
|---|---|---|----|---|
| y | −3 | −1.5 | −4.5 | 0 |

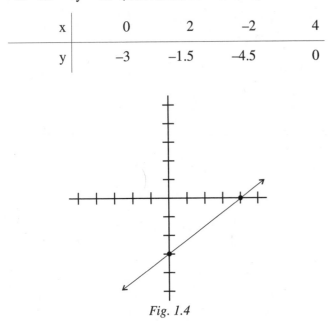

*Fig. 1.4*

4. $y = -\dfrac{3}{2}x - 4$ **(Cartesian coordinates)**

| x | 0 | 2 | –2 | –4 |
|---|---|---|----|----|
| y | –4 | –7 | –1 | 2 |

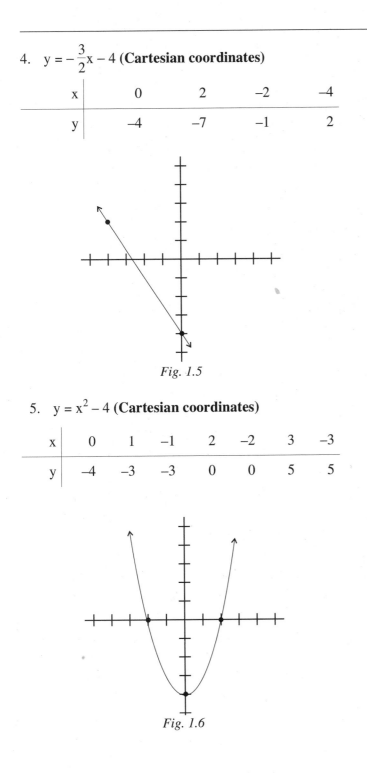

*Fig. 1.5*

5. $y = x^2 - 4$ **(Cartesian coordinates)**

| x | 0 | 1 | –1 | 2 | –2 | 3 | –3 |
|---|---|---|----|---|----|---|----|
| y | –4 | –3 | –3 | 0 | 0 | 5 | 5 |

*Fig. 1.6*

6.  $y = \sqrt{x + 1}$    (**Cartesian coordinates**)

| x | −1 | 0 | 3 | 8 |
|---|---|---|---|---|
| y | 0 | 1 | 2 | 3 |

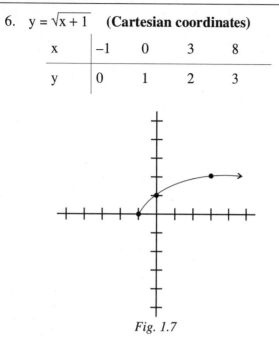

*Fig. 1.7*

7.  $x + y = 4$    (**Cartesian coordinates**)

| x | 0 | 2 | 3 | 4 |
|---|---|---|---|---|
| y | 4 | 2 | 1 | 0 |

$2x - y = 5$

| x | 0 | 2 | 3 | 4 |
|---|---|---|---|---|
| y | −5 | −1 | 1 | 3 |

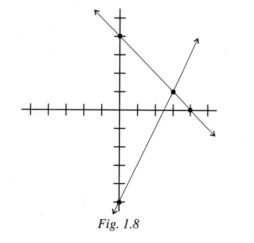

*Fig. 1.8*

8.  F = 1.8C + 32

    a. C = 30, so F = 1.8(30) + 32 = 54 + 32 = 86 degrees F.

    b. F = 50 = 1.8C + 32

    18 = 1.8C

    10 = C, or C = 10 degrees **(Cartesian coordinates)**

9.  2x + 5y = 10   **(Lines)**

| x | 5 | 0 | 2 |
|---|---|---|---|
| y | 0 | 2 | 1.2 |

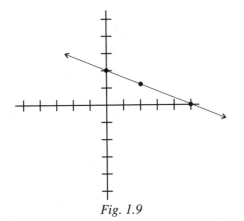

*Fig. 1.9*

10. Points are (−2, 3) and (−4, −5). $m = \dfrac{-5 - 3}{-4 - (-2)} = \dfrac{-8}{-2} = 4$ . **(Lines)**

11. Points are (3, −2) and (3, 4). $m = \dfrac{4 - (-2)}{3 - 3} = \dfrac{6}{0}$, which is undefined. **(Lines)**

12. Points are (4, 2) and (−5, 2). $m = \dfrac{2 - 2}{-5 - 4} = \dfrac{0}{-9} = 0$. **(Lines)**

13. (−1, −2) and (2, 3) are on the graph. $m = \dfrac{3 - (-2)}{2 - (-1)} = \dfrac{5}{3}$. **(Lines)**

14. m = −2, point (0, 3); thus, the y−intercept is 3.

    y = −2x + 3 **(Lines)**

15. m = 5, point (−1, 3); y − 3 = 5(x − (−1))

    y − 3 = 5(x + 1)

    y − 3 = 5x + 5

    y = 5x + 8 **(Lines)**

16. Points $(2, -7)$ and $(-1, -1)$. $m = \dfrac{-1-(-7)}{-1-2} = \dfrac{6}{-3} = -2$.

$$y-(-1) = -2(x-(-1))$$

$$y + 1 = -2x - 2$$

$$y = -2x - 3 \textbf{ (Lines)}$$

17. Points $(2, 2)$ and $(-7, 2)$. $m = \dfrac{2-2}{-7-2} = \dfrac{0}{-9} = 0$, $y = 2$. **(Lines)**

18. Point $(3, -6)$, parallel to $y = 4x - 7$. Parallel lines have the same slope, so $m = 4$.

$$y - (-6) = 4(x - 3)$$

$$y + 6 = 4x - 12$$

$$y = 4x - 18 \textbf{ (Lines)}$$

19. Point $(-2, 5)$, perpendicular to $y = -3x + 2$. $m = \dfrac{1}{3}$, since slopes of perpendicular lines have product $-1$.

$$y - 5 = \frac{1}{3}(x - (-2))$$

$$y - 5 = \frac{x}{3} + \frac{2}{3}$$

$$y = \frac{x}{3} + \frac{17}{3} \textbf{ (Lines)}$$

20. Points $(-2, 3)$ and $(1, 4)$ are on the graph. $m = \dfrac{4-3}{1-(-2)} = \dfrac{1}{3}$, so

$$y - 3 = \frac{1}{3}(x - (-2))$$

$$y - 3 = \frac{x}{3} + \frac{2}{3}$$

$$y = \frac{x}{3} + \frac{11}{3} \textbf{ (Lines)}$$

21. Point $(2, 3)$, perpendicular to $y = 17$, a horizontal line. Thus, the line sought is the vertical line through $(2, 3)$. The equation is $x = 2$. **(Lines)**

22. $6y - 2x = 12$

$$6y = 2x + 12$$

$$y = \frac{x}{3} + 2, \text{ so that } m = \frac{1}{3} \textbf{ (Lines)}$$

23. Point $(5, -2)$, parallel to $2x - 3y = 6$

Find the slope of $2x - 3y = 6$: $-3y = -2x + 6$, $y = \dfrac{2}{3}x - 2$, $m = \dfrac{2}{3}$

Parallel lines have the same slope. Thus,

$$y - (-2) = \frac{2}{3}(x - 5)$$

$$y + 2 = \frac{2x}{3} - \frac{10}{3}$$

$$y = \frac{2x}{3} - \frac{16}{3} \quad \textbf{(Lines)}$$

24. Point (4, –6), perpendicular to $3x + 5y = 15$

Find the slope of $3x + 5y = 15$: $5y = -3x + 15$, $y = \frac{-3x}{5} + 3$, $m = -\frac{3}{5}$. The slope of the perpendicular line is $m = \frac{5}{3}$. Thus,

$$y - (-6) = \frac{5}{3}(x - 4)$$

$$y + 6 = \frac{5x}{3} - \frac{20}{3}$$

$$y = \frac{5x}{3} - \frac{38}{3} \quad \textbf{(Lines)}$$

25. The cost of making x widgets is $C = 32x + 2000$ per month.

   a.   $x = 800$ widgets, the cost is $C = 32(800) + 2000 = 25600 + 2000 = 27{,}600$ dollars.

   b.   $C = 26{,}000$ dollars. Find x. $26000 = 32x + 2000$

$$24000 = 32x$$

$$750 = x \quad \textbf{(Lines)}$$

26. $f(x) = 2x + 10$

   a. $f(0) = 2(0) + 10 = 10$

   b. $f(-3) = 2(-3) + 10 = -6 + 10 = 4$

   c. $f(4) = 2(4) + 10 = 8 + 10 = 18$

   d. $f(a) = 2a + 10$

   e. $f(a + h) = 2(a + h) + 10 = 2a + 2h + 10$

   f. $f(2b - 1) = 2(2b - 1) + 10 = 4b - 2 + 10 = 4b + 8$ **(Functions)**

27. $g(t) = \frac{t(t - 1)}{t + 1}$.

   a. $g(0) = \frac{0(0 - 1)}{0 + 1} = \frac{0}{1} = 0$

   b. $g(1) = \frac{1(1 - 1)}{1 + 1} = \frac{0}{2} = 0$

   c. $g(2) = \frac{2(2 - 1)}{2 + 1} = \frac{2}{3}$

   d. $g(2c) = \frac{2c(2c - 1)}{2c + 1} = \frac{4c^2 - 2c}{2c + 1}$

e. $g(a + 1) = \dfrac{(a + 1)(a + 1 - 1)}{(a + 1) + 1} = \dfrac{a(a + 1)}{a + 2} = \dfrac{a^2 + a}{a + 2}$

f. $g(b) = \dfrac{b(b - 1)}{b + 1} = \dfrac{b^2 - b}{b + 1}$  (**Functions**)

28.  The values of F(x) are given by the following table.

| x | 0 | 1 | 2 | 3 | 4 | 5 | 6 |
|---|---|---|---|---|---|---|---|
| F(x) | 0 | 4 | 8 | 12 | 16 | 20 | 24 |

a. Find F(4) = 16.

b. Find F(6) = 24.

c. F(x) = 4x (**Functions**)

29.  A plumber charges $50.00 for a house call and $40.00 per hour. (**Functions**)

| hours | 1 | 2 | 3 | 4 | 5 |
|---|---|---|---|---|---|
| charge | $90 | $130 | $170 | $210 | $250 |

30.  x = the number of hours, C(x) = 40x +50 dollars per house call. (**Functions**)

31.  x = the number of souvenirs per month, C(x) = 2x + 1200 dollars. (**Functions**)

32.  R(x) = 5x dollars, where R = the revenue from selling x souvenirs. (**Functions**)

33.  x = the number of souvenirs, P(x) = revenue − cost = 5x − (2x + 1200) = 3x − 1200 dollars. (**Functions**)

34.  Break-even occurs when cost = revenue, or equivalently, when profit = 0. 0 = 3x − 1200, or 3x = 1200, or x = 400 souvenirs. (**Functions**)

35.  x = the number of miles driven, cost = C(x) = 25 + .05x dollars. (**Functions**)

36.  d = the number of the customer's video rental days per year, cost = C(d) = 2d dollars. (**Functions**)

37.  d = the number of video rental days per year, cost = V(d) = d + 10 dollars. (**Functions**)

38.  Points (0, 5000) and (5, 500). $m = \dfrac{5000 - 500}{0 - 5} = \dfrac{4500}{-5} = -900$. The y−intercept is 5000, so V(t) = 5000 − 900t. (**Functions**)

39.  a. V(2) = 5000 − 900(2) = 3200 dollars

b. V(3) = 5000 − 900(3) = 2300 dollars

c. V(4) = 5000 − 900(4) = 1400 dollars (**Functions**)

40.  The annual (straight line) depreciation for the copier in problems 38 and 39 is the absolute value of the slope of V = 5000 − 900t, or $900. (**Functions**)

# Grade Yourself

Circle the numbers of the questions you missed, then fill in the total incorrect for each topic. If you answered more than three questions incorrectly, you need to focus on that topic. (If a topic has less than three questions and you had at least one wrong, we suggest you study that topic also. Read your textbook or a review book, or ask your teacher for help.)

## Subject: Functions and Lines

| Topic | Question Numbers | Number Incorrect |
|---|---|---|
| Cartesian coordinates | 1, 2, 3, 4, 5, 6, 7, 8 | |
| Lines | 9, 10, 11, 12, 13, 14, 15, 16, 17, 18, 19, 20, 21, 22, 23, 24, 25 | |
| Functions | 26, 27, 28, 29, 30, 31, 32, 33, 34, 35, 36, 37, 38, 39, 40 | |

# *Matrices*

**2**

---

# Brief Yourself

An equation is linear if it has the form

$$a_1 x_1 + a_2 x_2 + ... + a_n x_n = b.$$

A system of linear equations is a set of m of these equations; to solve a system of equations is to find values for the variables $x_1$, $x_2$, ..., $x_n$ for which all the equations are satisfied. Finding the intersection of two lines is one example of solving a system of linear equations. The substitution method and the addition method are frequently used.

A second way to solve systems of linear equations is to use Gauss–Jordan reduction on the augmented matrix of the system.

Matrices can be added and subtracted if they have the same size. A matrix can also be multiplied by scalar, or constant. Two matrices can be multiplied if the number of columns of the left matrix equals the number of rows of the right matrix. Matrix multiplication is non-commutative. The transpose of a matrix is the matrix produced by interchanging the rows and columns of the original matrix.

Two square matrices are inverses of each other if their product equals the identity matrix. Gauss-Jordan elimination can be used to find the inverse of a square matrix, if it exists.

Leontief input-output models are a common application of matrices. In the closed model, everything produced by the economy is also consumed by the economy. In the open model, there is surplus produced for use outside the industries. We set up the input-output matrix using the mnemonic

$$
A = \text{TO} \begin{matrix} & & \text{FROM} \\ & & 1 \ 2 \\ \begin{matrix} 1 \\ 2 \end{matrix} & \begin{bmatrix} \phantom{xx} \\ \phantom{xx} \end{bmatrix} \end{matrix}
$$

The final demand on the economy is a matrix D. We solve an equation $X - AX = (I - A)X = D$ obtaining $X = (I - A)^{-1}D$ to find production levels necessary to meet demand.

# Test Yourself

1. Find the intersection of $3x + 4y = 12$ with $y = 5x - 2$ using the substitution method.

2. Find the intersection of $2x + 3y = 6$ with $3y = -2x + 15$ using the substitution method.

3. Find the intersection of $y = -2x + 5$ with $4x + 2y = 10$ using the substitution method.

4. Find the intersection of $2x + 7y = 1$ with $3x - 5y = 17$ using the substitution method.

5. Find the intersection of $-2x + y = 6$ with $4x - 2y = 10$ using the addition method.

6. Find the intersection of $3x - y = 2$ with $-6x + 2y + 4 = 0$ using the addition method.

7. Find the intersection of $2x + 5y = 22$ with $x - 2y = 2$ using the addition method.

8. Find the intersection of $2x + 4y = -4$ with $3x + 5y = -1$ using the addition method.

9. The demand equation for a certain commodity (the number of items, $x$, that customers will buy at price $y$) is $y = 2000 - 12x$. The supply equation (the number of items, $x$, that manufacturers are willing to supply at price $y$) is $y = 20x - 1200$. Find the equilibrium solution (price and number of items).

10. It costs $100 per month plus 50 cents an item for Brianna to manufacture party favors. If she sells them for 75 cents each, how many must she sell to break even?

11. Which of the following equations is linear?

    a. $4x = 6 - y$

    b. $y = \sqrt{x} + 2$

    c. $\sqrt{2}x + 3y = 1$

    d. $3x_2 + 4x_3 = x_1$

    e. $z = 4x^2 + 3y - 5$

    f. $y = \dfrac{1}{2x - 3}$

12. A system of 2 linear equations with 2 unknowns could have (write the letters of all possible answers)

    a. exactly 1 solution

    b. no solutions

    c. exactly 2 solutions

    d. infinitely many solutions

    e. exactly 4 solutions

    f. exactly 100 solutions

13. Write down the augmented matrix corresponding to the following system of linear equations.

    $2x - 4y = 3$

    $-x + 3y = 12$

14. Write down the augmented matrix corresponding to the following linear system.

    $4x - 7y = 3$

    $x + 2y = 4$

    $-2x - 5y = 11$

15. Which of the following matrices is in reduced row echelon form?

    a. $\left[\begin{array}{ccc|c} 1 & 2 & -3 & 2 \\ 0 & 0 & 1 & 4 \\ 0 & 0 & 0 & 1 \end{array}\right]$

    b. $\left[\begin{array}{ccccc} 1 & 0 & -3 & 1 & 2 \\ 0 & 0 & 0 & 0 & 0 \\ 0 & 1 & 0 & 1 & 3 \end{array}\right]$

    c. $\left[\begin{array}{ccccc} 1 & 2 & 2 & 0 & 5 \\ 0 & 1 & 3 & 0 & 4 \\ 0 & 0 & 0 & 1 & 3 \end{array}\right]$

    d. $\left[\begin{array}{ccc|c} 0 & 1 & 0 & 7 \\ 1 & 0 & 0 & 3 \\ 0 & 0 & 1 & 2 \end{array}\right]$

    e. $\left[\begin{array}{ccc|c} 1 & 0 & 0 & 2 \\ 0 & 1 & 0 & 8 \\ 0 & 0 & 3 & 6 \end{array}\right]$

f. $\begin{bmatrix} 1 & 0 & -1 & | & 0 \\ 0 & 1 & -2 & | & 0 \\ 0 & 0 & 0 & | & 0 \end{bmatrix}$

16. Solve using Gauss-Jordan elimination.

    $4x - 5y = 16$

    $5x + 3y = 20$

17. Solve using Gauss-Jordan elimination.

    $4x - 3y + 2z = 86$

    $x \quad\quad - 3z = -5$

    $2x + 2y + z = 1$

18. Solve using Gauss-Jordan elimination.

    $2x + 3y - 7z = 0$

    $-x - 4y + z = -7$

19. Solve using Gauss-Jordan elimination.

    $2x_1 + 3x_2 - 4x_3 - 4x_4 = 2$

    $4x_1 - 2x_2 + x_3 - 3x_4 = 0$

    $3x_2 - 2x_3 + 5x_4 = 6$

    $-2x_1 + x_2 \quad\quad + 4x_4 = 1$

20. A baker makes 3 types of desserts, cookies, cakes, and tortes. Each batch of cookies requires 4 cups of flour, 2 sticks of butter, and 3 eggs. Each cake requires 3 cups of flour, 1.5 sticks of butter, and 2 eggs, while each torte requires 5 cups of flour, 3 sticks of butter, and 4 eggs. If he has on hand 55 cups of flour, 30 sticks of butter, and 41 eggs, how many of each dessert can he make?

21. Given $A = \begin{bmatrix} 3 & 2 \\ -1 & 4 \\ 0 & 1 \end{bmatrix}$, $B = \begin{bmatrix} 2 & -1 & 0 \\ 5 & -3 & 4 \end{bmatrix}$,

    $C = \begin{bmatrix} 0 & 1 & -1 \\ -1 & 2 & 0 \\ 2 & 1 & 3 \end{bmatrix}$, $a = -2$ and $b = 4$, find,

    if possible,

    a. $aA$

    b. $C^t$

    c. $bA - aB^t$

    d. $AB$

    e. $AC$

    f. $C(aB^t)$

    g. $(aC)A$

    h. $(A + B)C$

*Note:* $A^t$ denotes the transpose of A; it is sometimes written $A'$.

22. Find two matrices $A \neq 0$ and $B \neq 0$ with $AB = BA$.

23. Find two matrices A and B with $AB \neq BA$.

24. Let $A = \begin{bmatrix} 1 & -1 \\ -2 & 2 \end{bmatrix}$ and $B = \begin{bmatrix} 3 & 2 \\ 3 & 2 \end{bmatrix}$.

    a. Find AB and BA.

    b. Is it possible for either A or B to have an inverse? Why or why not?

25. Let $A = \begin{bmatrix} 2 & 3 \\ 5 & 7 \end{bmatrix}$. Find $A^{-1}$ if it exists.

26. Let $A = \begin{bmatrix} 2 & 0 & 0 \\ 0 & -1 & 0 \\ 0 & 0 & 5 \end{bmatrix}$. Find $A^{-1}$ if it exists.

27. Let $A = \begin{bmatrix} 2 & 3 & 4 \\ 1 & 2 & 3 \\ -1 & -1 & -1 \end{bmatrix}$. Find $A^{-1}$ if it exists.

28. Let $A = \begin{bmatrix} 0 & 3 & 1 \\ 4 & 5 & 2 \\ -1 & 0 & 0 \end{bmatrix}$. Find $A^{-1}$ if it exists.

29. Let $A = \begin{bmatrix} 1 & 0 & 0 \\ 0 & 3 & -1 \\ 2 & 1 & -1 \end{bmatrix}$. Find A.

30. Solve using matrix inverses.

    $4x - 4y + z = -2$

    $-x \quad\quad - z = -3$

    $3y + 2z = 11$

31. A farmer and a tailor and a carpenter consume each others' produce. Out of each $1.00 of food, the farmer consumes $.40, the tailor $.20, and the carpenter $.20. Out each $1.00 of clothing, the farmer uses $.24, the tailor uses $.12, and the carpenter uses $.12. Finally, out of each $1.00 of carpentry, the farmer uses $.09, the tailor uses $.17, and the carpenter uses $.42.

    a. Write down the input-output matrix.

    b. How much must each produce so that the farmer satisfies an external demand of $5,520 worth of food, the tailor a demand for $11,760 of clothing, and the carpenter a demand for $10,760 of woodworking?

32.  Suppose that the input-output matrix for a closed economy with industries I, II, and III is

$$A = \begin{bmatrix} .2 & .3 & .4 \\ .2 & .1 & .4 \\ .6 & .6 & .2 \end{bmatrix}.$$

Suppose that industry III produces $1,000,000. What output must each industry produce that the economy meets all its demands exactly?

33.  Three backyard farmers agree to divide their crops as follows. Farmer A will grow zucchini. He will give .3 of his crop to Farmer B and .3 of his crop to Farmer C. Farmer B will grow tomatoes. He will give .4 of his crop to Farmer A and .2 of his crop to Farmer C. Farmer C will grow snap peas. He will give .2 of his crop to farmer A and .3 of his crop to Farmer B. What prices should be assigned to their respective crops if the Leontief condition of a closed economy is to be satisfied and if the lowest-priced crop is priced at $100.00?

34.  Suppose that in the previous problem, Farmer A keeps .2 of his zucchini and has an external demand for $280 of zucchini; Farmer B keeps .3 of his tomatoes and has a demand for $360 of tomatoes; and Farmer C keeps .2 of his snap peas and has a demand for $280 of snap peas. How much must each produce?

---

## ✓ Check Yourself

1.  Substitute $5x - 2$ for y in $3x + 4y = 12$, to obtain

$$3x + 4(5x - 2) = 12$$
$$3x + 20x - 8 = 12$$
$$23x = 20$$

$x = \dfrac{20}{23}$, $y = 5x - 2 = 5\left(\dfrac{20}{23}\right) - 2 = \dfrac{100}{23} - \dfrac{46}{23} = \dfrac{54}{23}$. The solution is $\left(\dfrac{20}{23}, \dfrac{54}{23}\right)$. **(Linear systems)**

2.  Solve for one variable in one of the equations; say, solve for x in

$2x + 3y = 6$.
$$2x = 6 - 3y$$
$$x = 3 - 1.5y$$

Substitute in the other equation: $3y = -2x + 15$
$$3y = -2(3 - 1.5y) + 15$$
$$3y = -6 + 3y + 15$$

$0 = 9$, a contradiction. There is no solution. **(Linear systems)**

3.  Substitute $y = -2x + 5$ in $4x + 2y = 10$
$$4x + 2(-2x + 5) = 10$$
$$4x - 4x + 10 = 10$$

$0 = 0$, always true. Every pair $(x, -2x + 5)$ is a solution.

**(Linear systems)**

4. Solve for one letter in one equation. Solve for x in

$$2x + 7y = 1$$

$$2x = 1 - 7y$$

$$x = .5 - 3.5y.$$

Substitute in the other equation: $3x - 5y = 17$

$$3(.5 - 3.5y) - 5y = 17$$

$$1.5 - 10.5y - 5y = 17$$

$$-15.5y = 15.5$$

$$y = -1, \text{ so } x = .5 - 3.5(-1) = .5 + 3.5 = 4.$$

The solution is $(4, -1)$. **(Linear systems)**

5.  $-2x + \ y = 6$

$4x - 2y = 10$

———————

$-4x + 2y = 12$ (multiply equation 1 by 2)

$4x - 2y = 10$

———————

$0 = 22$, so there are no solutions. **(Linear systems)**

6.  $3x - \ y \quad = 2$

$-6x + 2y + 4 = 0$

———————

$3x - \ y \quad = 2$

$-6x + 2y \quad = -4$  (move 4 to the other side of equation 2)

———————

$3x - \ y \quad = 2$

$-3x + \ y \quad = -2$  (multiply equation 2 by .5)

———————

$0 = 0$

Since (equation 1) $y = 3x - 2$, all points of the form $(x, 3x - 2)$ are solutions. **(Linear systems)**

7.  $2x + 5y = 22$

$x - 2y = \ 2$

———————

$2x + 5y = 22$

$-2x + 4y = -4$ (multiple equation 2 by $-2$)

———————

$9y = 18$, so $y = 2$. Thus $x - 2(2) = 2$, or $x = 6$. The solution is $(6, 2)$. **(Linear systems)**

8.  $2x + 4y = -4$

    $3x + 5y = -1$

    ___

    $6x + 12y = -12$ (multiply equation 1 by 3)

    $-6x - 10y = 2$ (multiply equation 2 by $-2$)

    ___

    $2y = -10$

    $y = -5$. To find x, substitute: $2x + 4(-5) = -4$, $2x = 16$, $x = 8$.   The solution is $(8, -5)$.
    **(Linear systems)**

9.  $y = 2000 - 12x$

    $y = 20x - 1200$

    ___

    Substituting for y, $20x - 1200 = 2000 - 12x$

    $32x = 3200$

    $x = 100$ items at price $y = 2000 - 1200 = 800$ dollars. Equilibrium is $(100, 800)$. **(Linear systems)**

10. $C(x) = 100 + .5x$, and $R(x) = .75x$. Break-even occurs when $R(x) = C(x)$, or $.75x = 100 + .5x$

    $.25x = 100$

    $x = 400$ party favors **(Linear systems)**

11. Equations a, c, and d are linear. (Note: $\sqrt{2}$ is a number.) **(Linear systems)**

12. a, b, and d are correct. **(Linear systems)**

13. $\begin{bmatrix} 2 & -4 & | & 3 \\ -1 & 3 & | & 12 \end{bmatrix}$ **(Linear systems)**

14. $\begin{bmatrix} 4 & -7 & | & 3 \\ 1 & 2 & | & 4 \\ -2 & -5 & | & 11 \end{bmatrix}$ **(Linear systems)**

15. Matrix f is in reduced row echelon form. **(Gauss-Jordan elimination)**

16. $\begin{bmatrix} 4 & -5 & | & 16 \\ 5 & 3 & | & 20 \end{bmatrix} \rightarrow \begin{bmatrix} 1 & -1.25 & | & 4 \\ 5 & 3 & | & 20 \end{bmatrix} \rightarrow \begin{bmatrix} 1 & -1.25 & | & 4 \\ 0 & 9.25 & | & 0 \end{bmatrix} \rightarrow \begin{bmatrix} 1 & -1.25 & | & 4 \\ 0 & 1 & | & 0 \end{bmatrix} \rightarrow \begin{bmatrix} 1 & 0 & | & 4 \\ 0 & 1 & | & 0 \end{bmatrix}$

    thus $x = 4$ and $y = 0$, so $(4, 0)$ is the solution. **(Gauss-Jordan elimination)**

17. $\begin{bmatrix} 4 & -3 & 2 & | & 86 \\ 1 & 0 & -3 & | & -5 \\ 2 & 2 & 1 & | & 1 \end{bmatrix} \rightarrow \begin{bmatrix} 1 & 0 & -3 & | & -5 \\ 4 & -3 & 2 & | & 86 \\ 2 & 2 & 1 & | & 1 \end{bmatrix} \rightarrow \begin{bmatrix} 1 & 0 & -3 & | & -5 \\ 0 & -3 & 14 & | & 106 \\ 0 & 2 & 7 & | & 11 \end{bmatrix} \rightarrow \begin{bmatrix} 1 & 0 & -3 & | & -5 \\ 0 & 2 & 7 & | & 11 \\ 0 & -3 & 14 & | & 106 \end{bmatrix} \rightarrow$

    $\begin{bmatrix} 1 & 0 & -3 & | & -5 \\ 0 & 1 & 3.5 & | & 5.5 \\ 0 & -3 & 14 & | & 106 \end{bmatrix} \rightarrow \begin{bmatrix} 1 & 0 & -3 & | & -5 \\ 0 & 1 & 3.5 & | & 5.5 \\ 0 & 0 & 24.5 & | & 122.5 \end{bmatrix} \rightarrow \begin{bmatrix} 1 & 0 & -3 & | & -5 \\ 0 & 1 & 3.5 & | & 5.5 \\ 0 & 0 & 1 & | & 5 \end{bmatrix} \rightarrow \begin{bmatrix} 1 & 0 & 0 & | & 10 \\ 0 & 1 & 0 & | & -12 \\ 0 & 0 & 1 & | & 5 \end{bmatrix}$

    or $x = 10$, $y = -12$, $z = 5$. The solution is $(10, -12, 5)$. **(Gauss-Jordan elimination)**

18. $\begin{bmatrix} 2 & 3 & -7 & | & 0 \\ -1 & -4 & 1 & | & -7 \end{bmatrix} \rightarrow \begin{bmatrix} 1 & 4 & -1 & | & 7 \\ 2 & 3 & -7 & | & 0 \end{bmatrix} \rightarrow \begin{bmatrix} 1 & 4 & -1 & | & 7 \\ 0 & -5 & -5 & | & -14 \end{bmatrix}$

$\begin{bmatrix} 1 & 4 & -1 & | & 7 \\ 0 & 1 & 1 & | & 2.8 \end{bmatrix} \rightarrow \begin{bmatrix} 1 & 0 & -5 & | & -4.2 \\ 0 & 1 & 1 & | & 2.8 \end{bmatrix}$, or x - 5z = -4.2 and y + z = 2.8. Thus, using parameters,

x = -4.2 + 5t

y = 2.8 - t

z = t          There are infinitely many solutions. **(Gauss-Jordan elimination)**

19. $\begin{bmatrix} 2 & 3 & -4 & -4 & | & 2 \\ 4 & -2 & 1 & -3 & | & 0 \\ 0 & 3 & -2 & 5 & | & 6 \\ -2 & 1 & 0 & 4 & | & 1 \end{bmatrix} \rightarrow \begin{bmatrix} 2 & 3 & -4 & -4 & | & 2 \\ 0 & -8 & 9 & 5 & | & -4 \\ 0 & 3 & -2 & 5 & | & 6 \\ 0 & 4 & -4 & 0 & | & 3 \end{bmatrix} \rightarrow \begin{bmatrix} 2 & 3 & -4 & -4 & | & 2 \\ 0 & 4 & -4 & 0 & | & 3 \\ 0 & 3 & -2 & 5 & | & 6 \\ 0 & -8 & 9 & 5 & | & -4 \end{bmatrix} \rightarrow$

$\begin{bmatrix} 2 & 3 & -4 & -4 & | & 2 \\ 0 & 1 & -1 & 0 & | & 0.75 \\ 0 & 3 & -2 & 5 & | & 6 \\ 0 & 0 & 1 & 5 & | & 2 \end{bmatrix} \rightarrow \begin{bmatrix} 2 & 3 & -4 & -4 & | & 2 \\ 0 & 1 & -1 & 0 & | & 0.75 \\ 0 & 0 & 1 & 5 & | & 3.75 \\ 0 & 0 & 1 & 5 & | & 2 \end{bmatrix} \rightarrow \begin{bmatrix} 2 & 3 & -4 & -4 & | & 2 \\ 0 & 1 & -1 & 0 & | & 0.75 \\ 0 & 0 & 1 & 5 & | & 3.75 \\ 0 & 0 & 0 & 0 & | & 1.75 \end{bmatrix}$

Since the last row has all zeros on the left, but a non-zero number on the right, the last equation is 0 = 1.75, a contradiction. Therefore, there is no solution. **(Gauss-Jordan elimination)**

20. Let x = the number of batches of cookies, y = the number of cakes, and z = the number of tortes.

Then,   4x +  3 y + 5z = 55 = the amount of flour

2x + 1.5y + 3z = 30 = the amount of butter

3x +  2y + 4z = 41 = the number of eggs

Thus, $\begin{bmatrix} 4 & 3 & 5 & | & 55 \\ 2 & 1.5 & 3 & | & 30 \\ 3 & 2 & 4 & | & 41 \end{bmatrix} \rightarrow \begin{bmatrix} 2 & 1.5 & 3 & | & 30 \\ 4 & 3 & 5 & | & 55 \\ 3 & 2 & 4 & | & 41 \end{bmatrix} \rightarrow \begin{bmatrix} 2 & 1.5 & 3 & | & 30 \\ 0 & 0 & -1 & | & -5 \\ 3 & 2 & 4 & | & 41 \end{bmatrix} \rightarrow \begin{bmatrix} 1 & 0.75 & 1.5 & | & 15 \\ 3 & 2 & 4 & | & 41 \\ 0 & 0 & 1 & | & 5 \end{bmatrix} \rightarrow$

$\begin{bmatrix} 1 & 0.75 & 1.5 & | & 15 \\ 0 & -0.25 & -0.5 & | & -4 \\ 0 & 0 & 1 & | & 5 \end{bmatrix} \rightarrow \begin{bmatrix} 1 & 0.75 & 1.5 & | & 15 \\ 0 & 1 & 2 & | & 16 \\ 0 & 0 & 1 & | & 5 \end{bmatrix} \rightarrow \begin{bmatrix} 1 & 0.75 & 0 & | & 7.5 \\ 0 & 1 & 0 & | & 6 \\ 0 & 0 & 1 & | & 5 \end{bmatrix} \rightarrow \begin{bmatrix} 1 & 0 & 0 & | & 3 \\ 0 & 1 & 0 & | & 6 \\ 0 & 0 & 1 & | & 5 \end{bmatrix}$

Thus, x = 3, y = 6, and z = 5. That is, he can make 3 batches of cookies, 6 cakes and 5 tortes. **(Gauss-Jordan elimination)**

21. a. $aA = -2\begin{bmatrix} 3 & 2 \\ -1 & 4 \\ 0 & 1 \end{bmatrix} = \begin{bmatrix} -6 & -4 \\ 2 & -8 \\ 0 & -2 \end{bmatrix}$

b. $C^t = \begin{bmatrix} 0 & -1 & 2 \\ 1 & 2 & 1 \\ -1 & 0 & 3 \end{bmatrix}$

c. $bA - aB^t = 4\begin{bmatrix} 3 & 2 \\ -1 & 4 \\ 0 & 1 \end{bmatrix} - (-2)\begin{bmatrix} 2 & -1 & 0 \\ 5 & -3 & 4 \end{bmatrix}^t = \begin{bmatrix} 12 & 8 \\ -4 & 16 \\ 0 & 4 \end{bmatrix} + 2\begin{bmatrix} 2 & 5 \\ -1 & -3 \\ 0 & 4 \end{bmatrix} = \begin{bmatrix} 12 & 8 \\ -4 & 16 \\ 0 & 4 \end{bmatrix} + \begin{bmatrix} 4 & 10 \\ -2 & -6 \\ 0 & 8 \end{bmatrix} = \begin{bmatrix} 16 & 18 \\ -6 & 10 \\ 0 & 12 \end{bmatrix}$

d. $AB = \begin{bmatrix} 3 & 2 \\ -1 & 4 \\ 0 & 1 \end{bmatrix} \begin{bmatrix} 2 & -1 & 0 \\ 5 & -3 & 4 \end{bmatrix}$

$\begin{bmatrix} 3(2) + 2(5) & 3(-1) + 2(-3) & 3(0) + 2(4) \\ -1(2) + 4(5) & -1(-1) + 4(-3) & -1(0) + 4(4) \\ 0(2) + 1(5) & 0(-1) + 1(-3) & 0(0) + 1(4) \end{bmatrix}$

$\begin{bmatrix} 16 & -9 & 8 \\ 18 & -11 & 16 \\ 5 & -3 & 4 \end{bmatrix}$

e. AC is undefined since the number of columns of A (2) is not equal to the number of rows of C (3).

f. $C(aB^t) = \begin{bmatrix} 0 & 1 & -1 \\ -1 & 2 & 0 \\ 2 & 1 & 3 \end{bmatrix} \left( -2 \begin{bmatrix} 2 & -1 & 0 \\ 5 & -3 & 4 \end{bmatrix}^t \right) = \begin{bmatrix} 0 & 1 & -1 \\ -1 & 2 & 0 \\ 2 & 1 & 3 \end{bmatrix} \begin{bmatrix} -4 & -10 \\ 2 & 6 \\ 0 & -8 \end{bmatrix}$

$= \begin{bmatrix} 2 & 14 \\ 8 & 22 \\ -6 & -38 \end{bmatrix}$

g. $(aC)A = \left( -2 \begin{bmatrix} 0 & 1 & -1 \\ -1 & 2 & 0 \\ 2 & 1 & 3 \end{bmatrix} \right) \begin{bmatrix} 3 & 2 \\ -1 & 4 \\ 0 & 1 \end{bmatrix} = \begin{bmatrix} 0 & -2 & 2 \\ 2 & -4 & 0 \\ -4 & -2 & -6 \end{bmatrix} \begin{bmatrix} 3 & 2 \\ -1 & 4 \\ 0 & 1 \end{bmatrix} = \begin{bmatrix} 2 & -6 \\ 10 & -12 \\ -10 & -22 \end{bmatrix}$

h. (A + B)C is undefined since A + B is undefined (they do not have the same dimensions). **(Matrix algebra)**

22. There are many possible answers. Since AI = IA = A for all square matrices A (where I is the identity matrix of the same size), an easy choice for one of the matrices is I. Thus, A = I and B = any matrix will do.

If $A = \begin{bmatrix} 1 & 0 \\ 0 & 1 \end{bmatrix}$, $B = \begin{bmatrix} 1 & 3 \\ 2 & 4 \end{bmatrix}$, then AB = IB = B and BA = BI = B. **(Matrix algebra)**

23. Here, we need to choose A and B with more care. However, if we avoid the identity, there are so many examples, it is hard not to find one.

For example, if $A = \begin{bmatrix} 0 & 1 \\ 1 & 0 \end{bmatrix}$ and $B = \begin{bmatrix} 1 & 3 \\ 2 & 4 \end{bmatrix}$, then $AB = \begin{bmatrix} 0 & 1 \\ 1 & 0 \end{bmatrix} \begin{bmatrix} 1 & 3 \\ 2 & 4 \end{bmatrix} = \begin{bmatrix} 2 & 4 \\ 1 & 3 \end{bmatrix}$ and

$BA = \begin{bmatrix} 1 & 3 \\ 2 & 4 \end{bmatrix} \begin{bmatrix} 0 & 1 \\ 1 & 0 \end{bmatrix} = \begin{bmatrix} 3 & 1 \\ 4 & 2 \end{bmatrix}$, so that AB ≠ BA. **(Matrix algebra)**

24. a. $AB = \begin{bmatrix} 1 & -1 \\ -2 & 2 \end{bmatrix} \begin{bmatrix} 3 & 2 \\ 3 & 2 \end{bmatrix} = \begin{bmatrix} 0 & 0 \\ 0 & 0 \end{bmatrix}$ and $BA = \begin{bmatrix} 3 & 2 \\ 3 & 2 \end{bmatrix} \begin{bmatrix} 1 & -1 \\ -2 & 2 \end{bmatrix} = \begin{bmatrix} -1 & 1 \\ -1 & 1 \end{bmatrix}$. **(Matrix algebra)**

b. Neither A nor B can have an inverse since A ≠ 0, B ≠ 0 and AB = 0. **(Matrix inverses)**

25. To find $A^{-1}$, reduce $\begin{bmatrix} 2 & 3 & | & 1 & 0 \\ 5 & 7 & | & 0 & 1 \end{bmatrix}$:

$\begin{bmatrix} 2 & 3 & | & 1 & 0 \\ 5 & 7 & | & 0 & 1 \end{bmatrix} \rightarrow \begin{bmatrix} 1 & 1.5 & | & .5 & 0 \\ 5 & 7 & | & 0 & 1 \end{bmatrix} \rightarrow \begin{bmatrix} 1 & 1.5 & | & .5 & 0 \\ 0 & -.5 & | & -2.5 & 1 \end{bmatrix} \rightarrow \begin{bmatrix} 1 & 1.5 & | & .5 & 0 \\ 0 & 1 & | & 5 & -2 \end{bmatrix} \rightarrow \begin{bmatrix} 1 & 0 & | & -7 & 3 \\ 0 & 1 & | & 5 & -2 \end{bmatrix}$

so $A^1 = \begin{bmatrix} -7 & 3 \\ 5 & -2 \end{bmatrix}$. To check, multiply:

$AA^{-1} = \begin{bmatrix} 2 & 3 \\ 5 & 7 \end{bmatrix}\begin{bmatrix} -7 & 3 \\ 5 & -2 \end{bmatrix} = \begin{bmatrix} 1 & 0 \\ 0 & 1 \end{bmatrix}$. **(Matrix inverses)**

26. To find $A^{-1}$, reduce $\left[\begin{array}{ccc|ccc} 2 & 0 & 0 & 1 & 0 & 0 \\ 0 & -1 & 0 & 0 & 1 & 0 \\ 0 & 0 & 5 & 0 & 0 & 1 \end{array}\right]$:

$\left[\begin{array}{ccc|ccc} 2 & 0 & 0 & 1 & 0 & 0 \\ 0 & -1 & 0 & 0 & 1 & 0 \\ 0 & 0 & 5 & 0 & 0 & 1 \end{array}\right] \rightarrow \left[\begin{array}{ccc|ccc} 1 & 0 & 0 & .5 & 0 & 0 \\ 0 & 1 & 0 & 0 & -1 & 0 \\ 0 & 0 & 1 & 0 & 0 & .2 \end{array}\right]$, so $A^{-1} = \begin{bmatrix} .5 & 0 & 0 \\ 0 & -1 & 0 \\ 0 & 0 & .2 \end{bmatrix}$. **(Matrix inverses)**

27. To find $A^{-1}$, reduce $\left[\begin{array}{ccc|ccc} 2 & 3 & 4 & 1 & 0 & 0 \\ 1 & 2 & 3 & 0 & 1 & 0 \\ -1 & -1 & -1 & 0 & 0 & 1 \end{array}\right]$:

$\left[\begin{array}{ccc|ccc} 2 & 3 & 4 & 1 & 0 & 0 \\ 1 & 2 & 3 & 0 & 1 & 0 \\ -1 & -1 & -1 & 0 & 0 & 1 \end{array}\right] \rightarrow \left[\begin{array}{ccc|ccc} 1 & 2 & 3 & 0 & 1 & 0 \\ 2 & 3 & 4 & 1 & 0 & 0 \\ -1 & -1 & -1 & 0 & 0 & 1 \end{array}\right] \rightarrow \left[\begin{array}{ccc|ccc} 1 & 2 & 3 & 0 & 1 & 0 \\ 0 & -1 & -2 & 1 & -2 & 0 \\ 0 & 1 & 2 & 0 & 1 & 1 \end{array}\right] \rightarrow$

$\left[\begin{array}{ccc|ccc} 1 & 2 & 3 & 0 & 1 & 0 \\ 0 & 1 & 2 & -1 & 2 & 0 \\ 0 & 0 & 0 & 1 & -1 & 1 \end{array}\right]$. We see by the last row that no inverse exists. **(Matrix inverses)**

28. To find $A^{-1}$, reduce $\left[\begin{array}{ccc|ccc} 0 & 3 & 1 & 1 & 0 & 0 \\ 4 & 5 & 2 & 0 & 1 & 0 \\ -1 & 0 & 0 & 0 & 0 & 1 \end{array}\right]$:

$\left[\begin{array}{ccc|ccc} 0 & 3 & 1 & 1 & 0 & 0 \\ 4 & 5 & 2 & 0 & 1 & 0 \\ -1 & 0 & 0 & 0 & 0 & 1 \end{array}\right] \rightarrow \left[\begin{array}{ccc|ccc} 1 & 0 & 0 & 0 & 0 & -1 \\ 4 & 5 & 2 & 0 & 1 & 0 \\ 0 & 3 & 1 & 1 & 0 & 0 \end{array}\right] \rightarrow \left[\begin{array}{ccc|ccc} 1 & 0 & 0 & 0 & 0 & -1 \\ 0 & 5 & 2 & 0 & 1 & 4 \\ 0 & 3 & 1 & 1 & 0 & 0 \end{array}\right] \rightarrow$

$\left[\begin{array}{ccc|ccc} 1 & 0 & 0 & 0 & 0 & -1 \\ 0 & 2 & 1 & -1 & 1 & 4 \\ 0 & 3 & 1 & 1 & 0 & 0 \end{array}\right] \rightarrow \left[\begin{array}{ccc|ccc} 1 & 0 & 0 & 0 & 0 & -1 \\ 0 & 1 & .5 & -.5 & .5 & 2 \\ 0 & 3 & 1 & 1 & 0 & 0 \end{array}\right] \rightarrow \left[\begin{array}{ccc|ccc} 1 & 0 & 0 & 0 & 0 & -1 \\ 0 & 1 & .5 & -.5 & .5 & 2 \\ 0 & 0 & -.5 & 2.5 & -1.5 & -6 \end{array}\right] \rightarrow$

$\left[\begin{array}{ccc|ccc} 1 & 0 & 0 & 0 & 0 & -1 \\ 0 & 1 & 0 & 2 & -1 & -4 \\ 0 & 0 & 1 & -5 & 3 & 12 \end{array}\right]$. Thus, $A^{-1} = \begin{bmatrix} 0 & 0 & -1 \\ 2 & -1 & -4 \\ -5 & 3 & 12 \end{bmatrix}$. **(Matrix inverses)**

29. Let $A^{-1} = \begin{bmatrix} 1 & 0 & 0 \\ 0 & 3 & -1 \\ 2 & 1 & -1 \end{bmatrix}$. Find A. Since $(A^{-1})^{-1} = A$, reduce

$\left[\begin{array}{ccc|ccc} 1 & 0 & 0 & 1 & 0 & 0 \\ 0 & 3 & -1 & 0 & 1 & 0 \\ 2 & 1 & -1 & 0 & 0 & 1 \end{array}\right] \rightarrow \left[\begin{array}{ccc|ccc} 1 & 0 & 0 & 1 & 0 & 0 \\ 0 & 3 & -1 & 0 & 1 & 0 \\ 0 & 1 & -1 & -2 & 0 & 1 \end{array}\right] \rightarrow \left[\begin{array}{ccc|ccc} 1 & 0 & 0 & 1 & 0 & 0 \\ 0 & 1 & -1 & -2 & 0 & 1 \\ 0 & 3 & -1 & 0 & 1 & 0 \end{array}\right] \rightarrow$

$\left[\begin{array}{ccc|ccc} 1 & 0 & 0 & 1 & 0 & 0 \\ 0 & 1 & -1 & -2 & 0 & 1 \\ 0 & 0 & 2 & 6 & 1 & -3 \end{array}\right] \rightarrow \left[\begin{array}{ccc|ccc} 1 & 0 & 0 & 1 & 0 & 0 \\ 0 & 1 & -1 & -2 & 0 & 1 \\ 0 & 0 & 1 & 3 & .5 & -1.5 \end{array}\right] \rightarrow$

$$\begin{bmatrix} 1 & 0 & 0 & | & 1 & 0 & 0 \\ 0 & 1 & 0 & | & 1 & .5 & -.5 \\ 0 & 0 & 1 & | & 3 & .5 & -1.5 \end{bmatrix} \quad A = \begin{bmatrix} 1 & 0 & 0 \\ 1 & .5 & -.5 \\ 3 & .5 & -1.5 \end{bmatrix}. \textbf{ (Matrix inverses)}$$

30.  The coefficient matrix is $A = \begin{bmatrix} 4 & -4 & 1 \\ -1 & 0 & -1 \\ 0 & 3 & 2 \end{bmatrix}$. To find $A^{-1}$, reduce

$$\begin{bmatrix} 4 & -4 & 1 & | & 1 & 0 & 0 \\ -1 & 0 & -1 & | & 0 & 1 & 0 \\ 0 & 3 & 2 & | & 0 & 0 & 1 \end{bmatrix} \rightarrow \begin{bmatrix} 1 & 0 & 1 & | & 0 & -1 & 0 \\ 4 & -4 & 1 & | & 1 & 0 & 0 \\ 0 & 3 & 2 & | & 0 & 0 & 1 \end{bmatrix} \rightarrow \begin{bmatrix} 1 & 0 & 1 & | & 0 & -1 & 0 \\ 0 & -4 & -3 & | & 1 & 4 & 0 \\ 0 & 3 & 2 & | & 0 & 0 & 1 \end{bmatrix} \rightarrow$$

$$\begin{bmatrix} 1 & 0 & 1 & | & 0 & -1 & 0 \\ 0 & 1 & 1 & | & -1 & -4 & -1 \\ 0 & 0 & -1 & | & 3 & 12 & 4 \end{bmatrix} \rightarrow \begin{bmatrix} 1 & 0 & 0 & | & 3 & 11 & 4 \\ 0 & 1 & 0 & | & 2 & 8 & 3 \\ 0 & 0 & 1 & | & -3 & -12 & -4 \end{bmatrix}.$$

Since $A^{-1} = \begin{bmatrix} 3 & 11 & 4 \\ 2 & 8 & 3 \\ -3 & -12 & -4 \end{bmatrix}$, the solution is $\begin{bmatrix} x \\ y \\ z \end{bmatrix} = A^{-1} \begin{bmatrix} -2 \\ -3 \\ 11 \end{bmatrix} = \begin{bmatrix} 3 & 11 & 4 \\ 2 & 8 & 3 \\ -3 & -12 & -4 \end{bmatrix} \begin{bmatrix} -2 \\ -3 \\ 11 \end{bmatrix} = \begin{bmatrix} 5 \\ 5 \\ -2 \end{bmatrix}.$

**(Matrix inverses)**

31.  a. The input–output matrix is $A = \begin{array}{c} \\ \text{TO} \end{array} \begin{array}{c} \\ \begin{array}{c} F \\ T \\ C \end{array} \end{array} \overset{\begin{array}{ccc} \text{FROM} \\ F \quad T \quad C \end{array}}{\begin{bmatrix} .40 & .24 & .09 \\ .20 & .12 & .17 \\ .20 & .12 & .42 \end{bmatrix}}$

b. The production, $\begin{bmatrix} x \\ y \\ z \end{bmatrix}$, can be found by reducing

$$\begin{bmatrix} .60 & -.24 & -.09 & | & 5520 \\ -.20 & .88 & -.17 & | & 11760 \\ -.20 & -.12 & .58 & | & 10760 \end{bmatrix} \rightarrow \begin{bmatrix} 1 & -40 & -.15 & | & 9200 \\ 0 & .80 & -.20 & | & 13600 \\ 0 & -.20 & .55 & | & 12600 \end{bmatrix} \rightarrow$$

$$\begin{bmatrix} 1 & 0 & -.25 & | & 16000 \\ 0 & 1 & -.25 & | & 17000 \\ 0 & 0 & .5 & | & 16000 \end{bmatrix} \rightarrow \begin{bmatrix} 1 & 0 & 0 & | & 24000 \\ 0 & 1 & 0 & | & 25000 \\ 0 & 0 & 1 & | & 32000 \end{bmatrix}.$$ Thus, the farmer should produce \$24,000,

the tailor, \$25,000, and the carpenter \$32,000. **(Leontief input-output)**

32.  Solve $AX = X$, or $(I - A)X = 0$. To do this, reduce

$$\begin{bmatrix} .8 & -.3 & -.4 \\ -.2 & .9 & -.4 \\ -.6 & -.6 & .8 \end{bmatrix} \rightarrow \begin{bmatrix} 1 & -.375 & -.5 \\ 0 & .825 & -.5 \\ 0 & -.825 & .5 \end{bmatrix} \rightarrow \begin{bmatrix} 1 & 0 & -.727 \\ 0 & 1 & -.606 \\ 0 & 0 & 0 \end{bmatrix} \text{ so that the production} = \begin{bmatrix} x \\ y \\ z \end{bmatrix} = \begin{bmatrix} .727t \\ .606t \\ t \end{bmatrix}.$$

Thus, if Industry I produces \$727,000, Industry II \$606,000, and Industry III \$1,000,000, the economy will be in equilibrium. **(Leontief input-output)**

33. The input-output matrix is $A = \begin{bmatrix} .4 & .4 & .2 \\ .3 & .4 & .3 \\ .3 & .2 & .5 \end{bmatrix}$. Solve $(I - A)X = 0$.

$$\begin{bmatrix} .6 & -.4 & -.2 \\ -.3 & .6 & -.3 \\ -.3 & -.2 & .5 \end{bmatrix} \rightarrow \begin{bmatrix} .6 & -.4 & -.2 \\ 0 & .4 & -.4 \\ 0 & -.4 & .4 \end{bmatrix} \rightarrow \begin{bmatrix} .3 & -.2 & -.1 \\ 0 & .4 & -.4 \\ 0 & 0 & 0 \end{bmatrix} \rightarrow \begin{bmatrix} .3 & 0 & -.3 \\ 0 & 1 & -1 \\ 0 & 0 & 0 \end{bmatrix} \rightarrow \begin{bmatrix} 1 & 0 & -1 \\ 0 & 1 & -1 \\ 0 & 0 & 0 \end{bmatrix},$$

so that $= \begin{bmatrix} x \\ y \\ z \end{bmatrix} = \begin{bmatrix} t \\ t \\ t \end{bmatrix}$. Each must grow $100 worth. **(Leontief input-output)**

34. The input-output matrix is $A = \begin{bmatrix} .2 & .4 & .2 \\ .3 & .3 & .3 \\ .3 & .2 & .2 \end{bmatrix}$, so that we must reduce $\begin{bmatrix} .8 & -.4 & -.2 & | & 280 \\ -.3 & .7 & -.3 & | & 360 \\ -.3 & -.2 & .8 & | & 280 \end{bmatrix} \rightarrow$

$$\begin{bmatrix} 1 & -.5 & -.25 & | & 350 \\ 0 & .55 & -.375 & | & 465 \\ 0 & -.35 & .725 & | & 385 \end{bmatrix} \rightarrow \begin{bmatrix} 1 & 0 & -.591 & | & 772.7273 \\ 0 & 1 & -.682 & | & 845.4545 \\ 0 & 0 & .486 & | & 688.9091 \end{bmatrix} \rightarrow \begin{bmatrix} 1 & 0 & 0 & | & 1600 \\ 0 & 1 & 0 & | & 1800 \\ 0 & 0 & 1 & | & 1400 \end{bmatrix},$$

so that Farmer A should produce $1,600 of zucchini, Farmer B should produce $1,800 of tomatoes, and Farmer C should produce $1,400 of snap peas. **(Leontief input-output)**

# Grade Yourself

Circle the numbers of the questions you missed, then fill in the total incorrect for each topic. If you answered more than three questions incorrectly, you need to focus on that topic. (If a topic has less than three questions and you had at least one wrong, we suggest you study that topic also. Read your textbook or a review book, or ask your teacher for help.)

## Subject: Functions and Lines

| Topic | Question Numbers | Number Incorrect |
|---|---|---|
| Linear systems | 1, 2, 3, 4, 5, 6, 7, 8, 9, 10, 11, 12, 13, 14 | |
| Gauss–Jordan elimination | 15, 16, 17, 18, 19, 20 | |
| Matrix algebra | 21, 22, 23, 24a | |
| Matrix inverses | 24b, 25, 26, 27, 28, 29, 30 | |
| Leontief input–output | 31, 32, 33, 34 | |

# Linear Programming

## 3

## Brief Yourself

A linear programming problem involves a linear function, the objective function, which is to be maximized or minimized, subject to constraints. The constraints are linear inequalities together with an assumption that all variables be non-negative.

The constraints give rise to a feasible region of solutions. Among these are some that maximize or minimize the objective function; the maximum or minimum occurs at a vertex, or at all points along a segment between two adjacent vertices. The geometric method can be used to solve problems whose constraints can be stated in terms of two variables. One finds the vertices of the feasible region and evaluates the objective function at each. If the feasible region is bounded, the smallest value of the objective function is the minimum and the largest is the maximum. If the feasible region is unbounded there may not be a maximum or a minimum.

The standard linear programming problem is to maximize a linear objective function $z = CX$, where $C$ is a row matrix, and $X$ is a column matrix of variables subject to the constraints $X \geq 0$ and $AX \leq B^t$. The dual problem is to minimize $z = B\,U$ subject to the constraints $U \geq 0$ and $A^t \geq U\,C^t$.

The algorithm used to solve the standard linear programming problem is the simplex method. First, slack variables are introduced to make $AX \leq B$ a system of linear equalities. The tableau is set up, and the pivot element is chosen: (1) Select the column containing the most negative entry in the bottom row. The column belongs to the entering variable. (2) Divide each entry of the last column by the entry of the pivot column. The row containing the smallest quotient belongs to the departing variable. If there are no positive entries in the column, the problem has no solution. The pivot entry is in the row and column selected. Using elementary row operations, make the pivot entry 1 and place 0s above and below it in the column. Continue as long as possible. When there are no negative entries in the bottom row, the maximal value of $z$ has been found. The solutions are the values in the last column corresponding to the variables with a 1 in the row and 0s above and below. The solution of the dual problem lies in the bottom row under the slack variables.

# Test Yourself

1. Solve and graph the inequalities:

   a. $4x - 5y \leq 10$

   b. $2x + 3y > 24$

2. Find the feasible region and solve using the geometric method.

   Maximize $M = 3x + 8y$

   subject to $x + y \geq 8$

   $$2x + 4y \leq 20$$

   $$x \geq 0, y \geq 0$$

3. Find the feasible region and solve using the geometric method.

   Minimize $m = 4x + y$

   subject to $x + 2y \geq 10$

   $$2x + y \geq 10$$

   $$x \geq 0, y \geq 0$$

4. Kate's Sub Shop sells regular subs made of 6 inches of bread and 2 ounces of meatballs, and large subs made of 10 inches of bread and 6 ounces of meatballs. The profit on a regular sub is $.75 and on a large sub is $1.25. If there are 120 feet of deli bread and 48 pounds of meatballs available, how many of each kind will maximize profit?

   a. Write the problem as a linear programming problem, identifying the objective function, the constraints, and the variables.

   b. Find the feasible region.

   c. Solve using the geometric method.

5. Lonnie is trying to minimize the number of calories he eats, but he must get enough nutrients. Food A has 4 gm of nutrient 1, 3 gm of nutrient 2, 8 gm of nutrient 3, and 200 calories per serving. Food B has 2 gm of nutrient 1, 6 gm of nutrient 2, 2 gm of nutrient 3, and 250 calories per serving. How much of each should he eat to receive at least 12 gm of nutrient 1, at least 18 gm of nutrient 2, and at least 16 gm of nutrient 3, and to minimize calories?

   a. Write the problem as a linear programming problem, identifying the objective function, the constraints, and the variables.

   b. Find the feasible region.

   c. Solve using the geometric method.

6. A shipper must ship some number of two types of items. Item A weighs 4 lb, and item B weighs 5 lb; the total weight should not exceed 30 lb. The recipient wants at least three items, and item A should not outnumber item B by more than 3. If the shipper makes $3 for each item A and $2 for each item B shipped, how many of each should be shipped to maximize profit?

   a. Write the problem as a linear programming problem, identifying the objective function, the constraints, and the variables.

   b. Find the feasible region.

   c. Solve using the geometric method.

   d. Does the objective function have a minimum also? If so, what is it?

7. Which of the following is a standard linear programming problem? How can you tell?

   a. Maximize $M = 3x - 2y + 5z$

   subject to $4x + 4y + z \leq 9$

   $$3x + 5y + 4z \leq 10$$

   $$x + 6y + 3z \leq 12$$

   b. Minimize $m = 5x + 4y + z$

   subject to $3x + 2y \leq 20$

   $$4x - z \leq 15$$

   $$x - 3y + 2z \leq 28$$

   $$x \geq 0, y \geq 0, z \geq 0$$

   c. Maximize $M = x + 6y$

   subject to $2x + 10y \leq 30$

   $$x - 4y \leq 25$$

   $$7x + 11y \leq 40$$

   $$x \geq 0, y \geq 0$$

   d. Maximize $M = 9x + 10y$

   subject to $5x + 6y \geq 3$

   $$2x + y \leq 9$$

   $$x \geq 0, y \geq 0$$

e. Minimize  $m = 5x + 4y + 8z$

subject to  $3x + 6y + 8z \geq 9$

$4x \qquad + 5z \geq 5$

$x \geq 0,\, y \geq 0,\, z \geq 0$

8.  In each of the following tableaux, select the next pivot entry.

a.
$$
\begin{bmatrix}
 & x & y & z & u & v & M & \text{sol} \\
\hline
u & -3 & 0 & 1 & 1 & 0 & 0 & 12 \\
y & 1 & 1 & 3 & 0 & 0 & 0 & 3 \\
v & 4 & 0 & -2 & 0 & 1 & 0 & 6 \\
\hline
M & -5 & 1 & -2 & 3 & 5 & 1 & 20
\end{bmatrix}
$$

b.
$$
\begin{bmatrix}
 & x & y & z & u & v & M & \text{sol} \\
\hline
x & 1 & 2 & -2 & 0 & 1 & 0 & 10 \\
u & 0 & 4 & 3 & 1 & -1 & 0 & 9 \\
\hline
M & 7 & -5 & -6 & 3 & 5 & 1 & 25
\end{bmatrix}
$$

9.  In each of the following tableaux, pivot on the indicated entry.

a.
$$
\begin{bmatrix}
 & x & y & z & u & v & M & \text{sol} \\
\hline
u & -3 & 0 & 1 & 1 & 0 & 0 & 12 \\
y & \boxed{1} & 1 & 3 & 0 & 0 & 0 & 3 \\
v & 4 & 0 & -2 & 0 & 1 & 0 & 6 \\
\hline
M & -5 & 1 & -2 & 3 & 5 & 1 & 20
\end{bmatrix}
$$

b.
$$
\begin{bmatrix}
 & x & y & z & u & v & M & \text{sol} \\
\hline
x & 1 & \boxed{2} & -2 & 0 & 1 & 0 & 10 \\
u & 0 & 4 & 3 & 1 & -1 & 0 & 9 \\
\hline
M & 7 & -5 & -6 & 3 & 5 & 1 & 25
\end{bmatrix}
$$

10.  Given the following standard linear programming problem:

Maximize  $M = 4x + 3y$

subject to  $3x + 5y \leq 15$

$2x + 3y \leq 12$

$x \geq 0,\, y \geq 0$

a. Solve using the simplex algorithm.

b. State the dual problem.

c. What is the solution of the dual problem?

11.  Given the following standard linear programming problem:

Maximize $M = 5x + 2y + 8z$

subject to  $x + 2y + \phantom{5}z \leq 20$

$4y + 5z \leq 30$

$2x + 3y \qquad \leq 12$

$x \geq 0,\, y \geq 0,\, z \geq 0$

a. Solve using the simplex algorithm.

b. State the dual problem.

c. What is the solution of the dual problem?

12.  Solve using the simplex algorithm. A computer store sells brand A, brand B, and brand C computers. The profit on the sale of each brand is $210, $140, and $70, respectively. The commissions are $20, $30, and $40, respectively, and the commission total should not exceed $3200. Finally, the different brands require warehouse space of 4, 4, and 5 cubic feet, respectively, and there is a total 480 cubic feet for storage. How many of each should be sold to maximize profits?

13.  Solve using the simplex algorithm.

Minimize  $m = 3x - 4y + z + 5$

subject to $x \qquad + 2z \geq 8$

$x + 3y + \phantom{3}z \leq 15$

$4y + 3z \leq 18$

$x \geq 0,\, y \geq 0,\, z \geq 0$

14.  Solve using the simplex algorithm.

Maximize  $M = 20x + 40y$

subject to  $x + \phantom{3}y \leq 60$

$2x + 3y \geq 20$

$x \geq 0,\, y \geq 0$

15.  Solve using the simplex algorithm.

Minimize  $m = 2x + 3y + 4z$

subject to  $3x + 5y \qquad \geq 60$

$x \qquad + z \geq 50$

$2y + z \geq 40$

$x \geq 0,\, y \geq 0,\, z \geq 0$

16. Solve using the simplex algorithm. Sondra's dog needs at least 25 gm of nutrient A, 50 gm of nutrient B. A serving of food 1 supplies 10 gm of nutrient A, 15 gm of nutrient B, and 300 calories, and costs 50 cents. A serving of food 2 supplies 12 gm of nutrient A, 20 gm of nutrient B, and 250 calories, and costs 40 cents. If the dog needs at least 1000 calories per day, how many serving of each food should he have to minimize cost?

# ✔ Check Yourself

1.  a. $4x - 5y \leq 10$

    First graph $4x - 5y = 10$ as a solid line, since the original inequality is $\leq$. Then choose a test point not on the line, such as $(0, 0)$. For $x = 0$, $y = 0$, the original inequality is true. Thus, the side containing $(0, 0)$ is the solution.

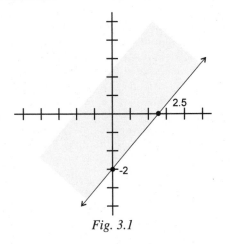

*Fig. 3.1*

b. $2x + 3y > 24$ First, graph $2x + 3y = 0$ as a dotted line, to indicate that the line is not part of the solution. Then, choose a test point, such as $(0, 0)$. For $x = 0$, $y = 0$, the inequality is false, so that the solution is the side not containing $(0, 0)$. **(Linear inequalities)**

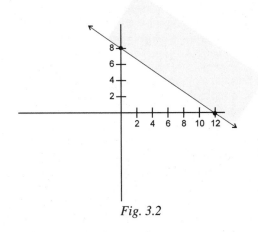

*Fig. 3.2*

2.  Maximize M = 3x + 8y

    subject to  x + y ≥ 8

    $\qquad$ 2x + 4y ≤ 20

    $\qquad$ x ≥ 0, y ≥ 0

    First, find the feasible region. Graph the lines x + y = 8 and 2x + 4y = 20.

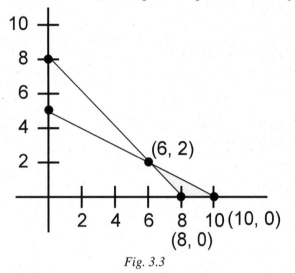

*Fig. 3.3*

Since x + y = 8

$\qquad$ 2x + 4y = 20

$\qquad$ –x – y = –8

$\qquad$ x + 2y = 10

$\qquad\qquad$ y = 2

Thus, x = 6, and the two lines intersect at (6, 2). The corner points are (6, 2), (8, 0), (10, 0).

| Corner point | M = 3x + 8y |
| --- | --- |
| (6, 2) | M = 18 + 16 = 34 |
| (8, 0) | M = 24 |
| (10, 0) | M = 30 |

So the maximal M = 34 for x = 6 and y = 2

**(Geometric solution)**

3.   Minimize   m = 4x + y

subject to   x + 2y ≥ 10

2x +   y ≥ 10

x ≥ 0, y ≥ 0

First, find the feasible region. Graph the lines x + 2y = 10 and 2x + y = 10.

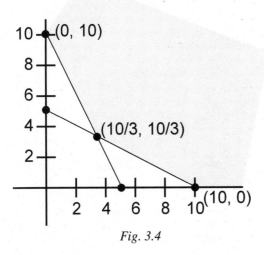

*Fig. 3.4*

Since x + 2y = 10

2x + y = 10

2x + 4y =  20

–2x –  y = –10

3y = 10

$y = \dfrac{10}{3}$. Then $x = \dfrac{10}{3}$, and the two lines intersect at (10/3, 10/3).

| Corner point | M = 4x + y |
|---|---|
| (10, 0) | M = 40 |
| (0, 10) | M = 10 |
| (10/3, 10/3) | M = 50/3 |

So the minimal m = 10 for x = 0 and y = 10.


**(Geometric solution)**

4.   a. Maximize M = .75x + 1.25y        (objective function)

subject to 6x + 10y ≤ 12(120) = 1440

   2x +  6y ≤ 16(48)  = 768

   x ≥ 0, y ≥ 0            (constraints)

x = the number of small subs, y = the number of large subs.

b. Graph the lines

   6x + 10y = 1440 and

   2x +  6y = 768

   –6x – 10y = –1440

   6x + 18y =  2304

         8y = 864,

y = 108, x = 60

The corner points are (60, 108), (240, 0), (0, 128), (0, 0).

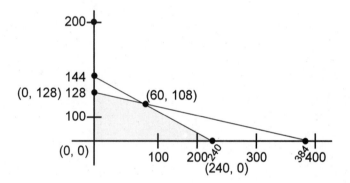

*Fig. 3.5*

c.

| Corner point | M = .75x + 1.25y |
|---|---|
| (60, 108) | M = 45 + 135 = 180 |
| (240, 0) | M = 180 |
| (0, 128) | M = 160 |
| (0, 0) | M = 0 |

The maximal profit is M = $180 for x = 60 small and
y = 108 large, and for x = 240 small and y = 0 large.

**(Geometric solution)**

5.  a. Minimize m = 200x + 250y        (objective function)

     subject to        $4x + 2y \geq 12$

                       $3x + 6y \geq 18$

                       $8x + 2y \geq 16$

                       $x \geq 0, y \geq 0$        (constraints)

x = the number of serving of food A, and y = the number of servings of food B.

b. Graph the lines

4x + 2y = 12, 3x + 6y = 18, and 8x + 2y = 16. The first two intersect at (2, 2); the first and third at (1, 4). The corner points are (2, 2), (1, 4), (6, 0), and (0, 8)

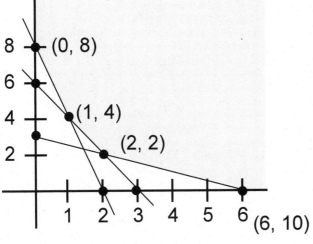

*Fig. 3.7*

c.

| Corner point | M = 200x + 250y |
|---|---|
| (2, 2) | M = 900 |
| (1, 4) | M = 1200 |
| (6, 0) | M = 1200 |
| (0, 8) | M = 2000 |

The minimal number of calories is 900 for x = 2 servings of food A and y = 2 servings of food B.

**(Geometric solution)**

6.   a. Maximize      M = 3x + 2y (objective function)

subject to 4x + 5y ≤ 30

x + y ≥ 3

x - y ≤ 3

x ≥ 0, y ≥ 0 (constraints)

x = the number of item A's shipped, and y = the number of item B's.

b. Graph the lines and find the corners. (3, 0), (5, 2), (0, 3), and (0, 6).

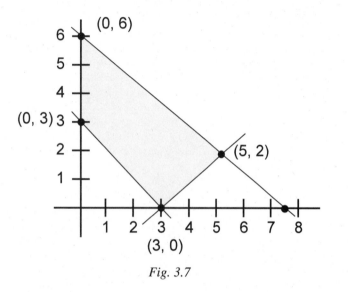

*Fig. 3.7*

c.

| Corner point | M = 3x + 2y |
|---|---|
| (3, 0) | M =  9 |
| (5, 2) | M = 19 |
| (0, 3) | M =  6 |
| (0, 6) | M = 12 |

The maximal profit is $19 for x = 5 of item A and
y = 2 of item B.

d. There is a minimal profit of M = $6 for x = 0 item A and y = 3 item B. **(Geometric solution)**

7.   a. Not a standard problem. The variables all need to be non-negative.

b. Not a standard problem. The objective function is not maximized.

c. Standard. The objective function is maximized, the variables are all positive, and the other constraints have the form linear expression ≤ number.

d. Not a standard problem. The first constraint is 5x + 6y ≥ 3, and it should be ≤ for a standard problem.

e. Not standard. The objective function should be maximized, and the inequalities in the linear constraints are wrong. This is a dual problem. **(Standard linear programming problem)**

8. a.

| | x | y | z | u | v | M | sol |
|---|---|---|---|---|---|---|---|
| u | -3 | 0 | 1 | 1 | 0 | 0 | 12 |
| y | 1 | 1 | 3 | 0 | 0 | 0 | 3 |
| v | [4] | 0 | -2 | 0 | 1 | 0 | 6 |
| M | -5 | 1 | -2 | 3 | 5 | 1 | 20 |

The "most negative" number in the last row is –5, in the x column. Thus, x is the entering variable. Next, examine 12/(–3), 3/1, 6/4. Of these the least positive quotient is 6/4, so the departing variable is v. The pivot is the 4 that is boxed.

b.

| | x | y | z | u | v | M | sol |
|---|---|---|---|---|---|---|---|
| x | 1 | 2 | -2 | 0 | 1 | 0 | 10 |
| u | 0 | 4 | [3] | 1 | -1 | 0 | 9 |
| M | 7 | -5 | -6 | 3 | 5 | 1 | 25 |

The "most negative" number in the last row is –6, in the z column. Thus, z is the entering variable. Next, examine 10/(–2), and 9/3. Of these the least positive quotient is 9/3, so the departing variable is u. The pivot is the 3 that is boxed. **(Simplex algorithm)**

9. a.

| | x | y | z | u | v | M | sol |
|---|---|---|---|---|---|---|---|
| u | -3 | 0 | 1 | 1 | 0 | 0 | 12 |
| y | [1] | 1 | 3 | 0 | 0 | 0 | 3 |
| v | 4 | 0 | -2 | 0 | 1 | 0 | 6 |
| M | -5 | 1 | -2 | 3 | 5 | 1 | 20 |

→

| | x | y | z | u | v | M | sol |
|---|---|---|---|---|---|---|---|
| u | 0 | 3 | 10 | 1 | 0 | 0 | 21 |
| y | [1] | 1 | 3 | 0 | 0 | 0 | 3 |
| v | 4 | 0 | -2 | 0 | 1 | 0 | 6 |
| M | -5 | 1 | -2 | 3 | 5 | 1 | 20 |

→

| | x | y | z | u | v | M | sol |
|---|---|---|---|---|---|---|---|
| u | 0 | 3 | 10 | 1 | 0 | 0 | 21 |
| y | [1] | 1 | 3 | 0 | 0 | 0 | 3 |
| v | 0 | -4 | -14 | 0 | 1 | 0 | -6 |
| M | -5 | 1 | -2 | 3 | 5 | 1 | 20 |

→

| | x | y | z | u | v | M | sol |
|---|---|---|---|---|---|---|---|
| u | 0 | 3 | 10 | 1 | 0 | 0 | 21 |
| y | [1] | 1 | 3 | 0 | 0 | 0 | 3 |
| v | 0 | -4 | -14 | 0 | 1 | 0 | -6 |
| M | 0 | 6 | -13 | 3 | 5 | 1 | 35 |

b.

| | x | y | z | u | v | M | sol |
|---|---|---|---|---|---|---|---|
| x | 1 | [2] | -2 | 0 | 1 | 0 | 10 |
| u | 0 | 4 | 3 | 1 | -1 | 0 | 9 |
| M | 7 | -5 | -6 | 3 | 5 | 1 | 25 |

→

| | x | y | z | u | v | M | sol |
|---|---|---|---|---|---|---|---|
| x | .5 | [1] | -1 | 0 | .5 | 0 | 5 |
| u | 0 | 4 | 3 | 1 | -1 | 0 | 9 |
| M | 7 | -5 | -6 | 3 | 5 | 1 | 25 |

→

| | x | y | z | u | v | M | sol |
|---|---|---|---|---|---|---|---|
| x | .5 | [1] | -1 | 0 | .5 | 0 | 5 |
| u | -2 | 0 | 7 | 1 | -3 | 0 | -11 |
| M | 7 | -5 | -6 | 3 | 5 | 1 | 25 |

→

| | x | y | z | u | v | M | sol |
|---|---|---|---|---|---|---|---|
| x | .5 | [1] | -1 | 0 | .5 | 0 | 5 |
| u | -2 | 0 | 7 | 1 | -3 | 0 | -11 |
| M | 9.5 | 0 | -11 | 3 | 7.5 | 1 | 50 |

**(Simplex algorithm)**

10. a. Introducing slack variables u and v, we have the equations

$$3x + 5y + u \qquad = 15$$
$$2x + 3y \qquad + v \qquad = 12$$
$$-4x - 3y \qquad\qquad + M = 0$$

The initial tableau is
$$\begin{bmatrix} & x & y & u & v & M & | & sol \\ u & \boxed{3} & 5 & 1 & 0 & 0 & | & 15 \\ v & 2 & 3 & 0 & 1 & 0 & | & 12 \\ - & - & - & - & - & - & | & - \\ M & -4 & -3 & 0 & 0 & 1 & | & 0 \end{bmatrix} \rightarrow \begin{bmatrix} & x & y & u & v & M & | & sol \\ x & \boxed{1} & \tfrac{5}{3} & \tfrac{1}{3} & 0 & 0 & | & 5 \\ v & 2 & 3 & 0 & 1 & 0 & | & 12 \\ - & - & - & - & - & - & | & - \\ M & -4 & -3 & 0 & 0 & 1 & | & 0 \end{bmatrix} \rightarrow$$

$$\begin{bmatrix} & x & y & u & v & M & | & sol \\ x & \boxed{1} & \tfrac{5}{3} & \tfrac{1}{3} & 0 & 0 & | & 5 \\ v & 0 & -\tfrac{1}{3} & -\tfrac{2}{3} & 1 & 0 & | & 2 \\ - & - & - & - & - & - & | & - \\ M & -4 & -3 & 0 & 0 & 1 & | & 0 \end{bmatrix} \rightarrow \begin{bmatrix} & x & y & u & v & M & | & sol \\ x & 1 & \tfrac{5}{3} & \tfrac{1}{3} & 0 & 0 & | & 5 \\ v & 0 & -\tfrac{1}{3} & -\tfrac{2}{3} & 1 & 0 & | & 2 \\ - & - & - & - & - & - & | & - \\ M & 0 & \tfrac{11}{3} & \tfrac{4}{3} & 0 & 1 & | & 20 \end{bmatrix}$$ . M = 20 for x = 5 and y = 0.

b. Since the original problem is maximize $M = [4 \; 3]\begin{bmatrix} x \\ y \end{bmatrix}$ subject to $\begin{bmatrix} 3 & 5 \\ 2 & 3 \end{bmatrix}\begin{bmatrix} x \\ y \end{bmatrix} \le \begin{bmatrix} 15 \\ 12 \end{bmatrix}, \begin{bmatrix} x \\ y \end{bmatrix} \ge 0$, the dual

problem is minimize $m = [15 \; 12]\begin{bmatrix} u \\ v \end{bmatrix} = 15u + 12v$ subject to $\begin{bmatrix} 3 & 2 \\ 5 & 3 \end{bmatrix}\begin{bmatrix} u \\ v \end{bmatrix} \ge \begin{bmatrix} 4 \\ 3 \end{bmatrix}$,

or $3u + 2v \ge 4$

$5u + 3v \ge 3$

$u \ge 0, v \ge 0$

c. m = 20 for u = 4/3 and v = 0.  (**Simplex method and dual problem**)

11. a. Introduce slack variables to obtain the equations

$$x + 2y + z + u \qquad\qquad = 20$$
$$4y + 5z + \qquad + v \qquad = 30$$
$$2x + 3y \qquad\qquad + w \quad = 12$$
$$-5x - 2y - 8z \qquad\qquad - M = 0$$

The initial tableau is
$$\begin{bmatrix} x & y & z & u & v & w & M & | & sol \\ 1 & 2 & \underline{1} & 1 & 0 & 0 & 0 & | & 20 \\ 0 & 4 & \boxed{5} & 0 & 1 & 0 & 0 & | & 30 \\ 2 & 3 & 0 & 0 & 0 & 1 & 0 & | & 12 \\ - & - & - & - & - & - & - & | & - \\ -5 & -2 & -8 & 0 & 0 & 0 & 1 & | & 0 \end{bmatrix} \rightarrow \begin{bmatrix} x & y & z & u & v & w & M & | & sol \\ 1 & 2 & \underline{1} & 1 & 0 & 0 & 0 & | & 20 \\ 0 & .8 & \boxed{1} & 0 & .2 & 0 & 0 & | & 6 \\ 2 & 3 & 0 & 0 & 0 & 1 & 0 & | & 12 \\ - & - & - & - & - & - & - & | & - \\ -5 & -2 & -8 & 0 & 0 & 0 & 1 & | & 0 \end{bmatrix} \rightarrow$$

$$\begin{bmatrix} x & y & z & u & v & w & M & | & sol \\ 1 & 1.2 & \underline{0} & 1 & -.2 & 0 & 0 & | & 14 \\ 0 & .8 & \boxed{1} & 0 & .2 & 0 & 0 & | & 6 \\ 2 & 3 & 0 & 0 & 0 & 1 & 0 & | & 12 \\ - & - & - & - & - & - & - & | & - \\ -5 & -2 & -8 & 0 & 0 & 0 & 1 & | & 0 \end{bmatrix} \rightarrow \begin{bmatrix} x & y & z & u & v & w & M & | & sol \\ 1 & 1.2 & \underline{0} & 1 & -.2 & 0 & 0 & | & 14 \\ 0 & .8 & \boxed{1} & 0 & .2 & 0 & 0 & | & 6 \\ 2 & 3 & 0 & 0 & 0 & 1 & 0 & | & 12 \\ - & - & - & - & - & - & - & | & - \\ -5 & 4.4 & 0 & 0 & 1.6 & 0 & 1 & | & 48 \end{bmatrix} \rightarrow \begin{bmatrix} x & y & z & u & v & w & M & | & sol \\ 1 & 1.2 & 0 & 1 & -.2 & 0 & 0 & | & 14 \\ 0 & .8 & 1 & 0 & .2 & 0 & 0 & | & 6 \\ \boxed{2} & 3 & 0 & 0 & 0 & 1 & 0 & | & 12 \\ - & - & - & - & - & - & - & | & - \\ -5 & 4.4 & 0 & 0 & 1.6 & 0 & 1 & | & 48 \end{bmatrix} \rightarrow$$

$$
\begin{bmatrix}
x & y & z & u & v & w & M & \text{sol} \\
1 & 1.2 & 0 & 1 & -.2 & 0 & 0 & 14 \\
0 & .8 & 1 & 0 & .2 & 0 & 0 & 6 \\
\boxed{1} & 1.5 & 0 & 0 & 0 & .5 & 0 & 6 \\
- & - & - & - & - & - & - & - \\
-5 & 4.4 & 0 & 0 & 1.6 & 0 & 1 & 48
\end{bmatrix}
\rightarrow
\begin{bmatrix}
x & y & z & u & v & w & M & \text{sol} \\
0 & -.3 & 0 & 1 & -.2 & -.5 & 0 & 8 \\
0 & .8 & 1 & 0 & .2 & 0 & 0 & 6 \\
\boxed{1} & 1.5 & 0 & 0 & 0 & .5 & 0 & 6 \\
- & - & - & - & - & - & - & - \\
0 & 11.9 & 0 & 0 & 1.6 & 2.5 & 1 & 78
\end{bmatrix}
$$

The maximal value of M is 78 when $x = 6$, $y = 0$, and $z = 6$.

b. Since the standard problem is: maximize $M = \begin{bmatrix} 5 & 2 & 8 \end{bmatrix} \begin{bmatrix} x \\ y \\ z \end{bmatrix}$

subject to $\begin{bmatrix} 1 & 2 & 1 \\ 0 & 4 & 5 \\ 2 & 3 & 0 \end{bmatrix} \begin{bmatrix} x \\ y \\ z \end{bmatrix} \le \begin{bmatrix} 20 \\ 30 \\ 12 \end{bmatrix}, \begin{bmatrix} x \\ y \\ z \end{bmatrix} \ge 0$, the dual problem is minimize $m = \begin{bmatrix} 20 & 30 & 12 \end{bmatrix} \begin{bmatrix} u \\ v \\ w \end{bmatrix} = 20u + 30v + 12w$

subject to $\begin{bmatrix} 1 & 0 & 2 \\ 2 & 4 & 3 \\ 1 & 5 & 0 \end{bmatrix} \begin{bmatrix} u \\ v \\ w \end{bmatrix} \ge \begin{bmatrix} 5 \\ 2 \\ 8 \end{bmatrix}, \begin{bmatrix} u \\ v \\ w \end{bmatrix} \ge 0$, or

$$u \qquad\quad + 2w \ge 5$$
$$2u + 4v \quad + 3w \ge 2$$
$$u + 5v \qquad\quad \ge 8$$
$$u \ge 0, v \ge 0, w \ge 0$$

c. The minimal value of $m = 78$ when $u = 0$, $v = 1.6$, and $w = 2.5$. (**Simplex method and dual problem**)

12. Maximize $M = 210x + 140y + 70z$

subject to $20x + 30y + 40z \le 3200$
$$4x + 4y + 5z \le 480$$
$$x \ge 0, y \ge 0, z \ge 0$$

Introduce the slack variables:
$$20x + 30y + 40z + u \qquad\qquad = 3200$$
$$4x + 4y + 5z \qquad + v \qquad = 480$$
$$-210x - 140y - 70z \qquad\qquad + M = 0$$

The initial tableau:
$$
\begin{bmatrix}
x & y & z & u & v & M & \text{sol} \\
20 & 30 & 40 & 1 & 0 & 0 & 3200 \\
\boxed{4} & 4 & 5 & 0 & 1 & 0 & 480 \\
- & - & - & - & - & - & - \\
-210 & -140 & -70 & 0 & 0 & 1 & 0
\end{bmatrix}
\rightarrow
\begin{bmatrix}
x & y & z & u & v & M & \text{sol} \\
20 & 30 & 40 & 1 & 0 & 0 & 3200 \\
\boxed{1} & 1 & 1.25 & 0 & .25 & 0 & 120 \\
- & - & - & - & - & - & - \\
-210 & -140 & -70 & 0 & 0 & 1 & 0
\end{bmatrix}
\rightarrow
$$

$$
\begin{bmatrix}
x & y & z & u & v & M & \text{sol} \\
0 & 10 & 15 & 1 & -5 & 0 & 800 \\
1 & 1 & 1.25 & 0 & .25 & 0 & 120 \\
- & - & - & - & - & - & - \\
0 & 70 & 192.5 & 0 & 52.5 & 1 & 25200
\end{bmatrix}
$$

The maximal profit of $25,200 is attained for $x = 120$ brand A computers, $y = 0$ brand B, and $z = 0$ brand C. (**Simplex algorithm**)

13. Maximize $M = -m = -3x + 4y - z - 5$

    subject to $-x \quad\quad -2z \le -8$

    $\quad\quad x + 3y + \ z \le 15$

    $\quad\quad\quad 4y + 3z \le 18$

    $\quad\quad x \ge 0, y \ge 0, z \ge 0$

Introduce the slack variables and set up the initial tableau.

$$\left[\begin{array}{ccccccc|c}
x & y & z & u & v & w & M & sol \\
-1 & 0 & \boxed{-2} & 1 & 0 & 0 & 0 & -8 \\
1 & 3 & 1 & 0 & 1 & 0 & 0 & 15 \\
0 & 4 & 3 & 0 & 0 & 1 & 0 & 18 \\
\hline
3 & -4 & 1 & 0 & 0 & 0 & 1 & -5
\end{array}\right] \rightarrow
\left[\begin{array}{ccccccc|c}
x & y & z & u & v & w & M & sol \\
.5 & 0 & \boxed{1} & -.5 & 0 & 0 & 0 & 4 \\
1 & 3 & 1 & 0 & 1 & 0 & 0 & 15 \\
0 & 4 & 3 & 0 & 0 & 1 & 0 & 18 \\
\hline
3 & -4 & 1 & 0 & 0 & 0 & 1 & -5
\end{array}\right] \rightarrow
\left[\begin{array}{ccccccc|c}
x & y & z & u & v & w & M & sol \\
.5 & 0 & 1 & -.5 & 0 & 0 & 0 & 4 \\
.5 & 3 & 0 & .5 & 1 & 0 & 0 & 11 \\
-1.5 & \boxed{4} & 0 & 1.5 & 0 & 1 & 0 & 6 \\
\hline
2.5 & -4 & 0 & .5 & 0 & 0 & 1 & -9
\end{array}\right] \rightarrow$$

$$\left[\begin{array}{ccccccc|c}
x & y & z & u & v & w & M & sol \\
.5 & 0 & 1 & -.5 & 0 & 0 & 0 & 4 \\
.5 & 3 & 0 & .5 & 1 & 0 & 0 & 11 \\
-\tfrac{3}{8} & \boxed{1} & 0 & \tfrac{3}{8} & 0 & .25 & 0 & 1.5 \\
\hline
2.5 & -4 & 0 & .5 & 0 & 0 & 1 & -9
\end{array}\right] \rightarrow
\left[\begin{array}{ccccccc|c}
x & y & z & u & v & w & M & sol \\
.5 & 0 & 1 & -.5 & 0 & 0 & 0 & -4 \\
1\tfrac{3}{8} & 0 & 0 & -\tfrac{5}{8} & 1 & -\tfrac{3}{4} & 0 & 1\tfrac{3}{2} \\
-\tfrac{3}{8} & \boxed{1} & 0 & \tfrac{3}{8} & 0 & \tfrac{1}{4} & 0 & \tfrac{3}{2} \\
\hline
1 & 0 & 0 & 2 & 0 & 0 & 1 & -3
\end{array}\right]$$

When $x = 0$, $y = 1.5$, and $z = 4$, $M = -3$, the maximum. Thus, at $x = 0$, $y = 1.5$, and $z = 4$, $m = 3x - 4y + z + 5$ has a minimum of $m = 3$. (**Nonstandard problems**)

14. Reverse the second constraint and introduce slack variables to obtain the equations

    $\quad x + \ y + u \quad\quad = 60$

    $-2x - \ 3y \quad + v \quad = -20$

    $-20x - 40y \quad\quad - M = \ 0$

The initial tableau is
$$\left[\begin{array}{ccccc|c}
x & y & u & v & M & sol \\
1 & \boxed{1} & 1 & 0 & 0 & 60 \\
-2 & -3 & 0 & 1 & 0 & -20 \\
\hline
-20 & -40 & 0 & 0 & 1 & 0
\end{array}\right] \rightarrow
\left[\begin{array}{ccccc|c}
x & y & u & v & M & sol \\
1 & 1 & 1 & 0 & 0 & 60 \\
1 & 0 & 3 & 1 & 0 & 160 \\
\hline
20 & 0 & 40 & 0 & 1 & 2400
\end{array}\right]$$

A maximum of $M = 2400$ is attained when $x = 0$ and $y = 60$. (**Nonstandard Problems**)

15. The original problem is minimize $m = [2\ 3\ 4]\begin{bmatrix} x \\ y \\ z \end{bmatrix}$

    subject to $\begin{bmatrix} 3 & 5 & 0 \\ 1 & 0 & 1 \\ 0 & 2 & 1 \end{bmatrix}\begin{bmatrix} x \\ y \\ z \end{bmatrix} \ge \begin{bmatrix} 60 \\ 50 \\ 40 \end{bmatrix}$, $\begin{bmatrix} x \\ y \\ z \end{bmatrix} \ge 0$. This is a dual problem.

    Solve maximize $M = [60\ 50\ 40]\begin{bmatrix} u \\ v \\ w \end{bmatrix} = 60u + 50v + 40w$

subject to $\begin{bmatrix} 3 & 1 & 0 \\ 5 & 0 & 2 \\ 0 & 1 & 1 \end{bmatrix} \begin{bmatrix} u \\ v \\ w \end{bmatrix} \leq \begin{bmatrix} 2 \\ 3 \\ 4 \end{bmatrix}, \begin{bmatrix} u \\ v \\ w \end{bmatrix} \geq 0$, that is

$$3u + v \qquad \leq 2$$

$$5u + \qquad 2w \leq 3$$

$$v + w \leq 4$$

$$u \geq 0, v \geq 0, w \geq 0$$

Introducing slack variables, the initial tableau is

$$\begin{bmatrix} u & v & w & r & s & t & M & | & sol \\ 3 & 1 & 0 & 1 & 0 & 0 & 0 & | & 2 \\ \boxed{5} & 0 & 2 & 0 & 1 & 0 & 0 & | & 3 \\ 0 & 1 & 1 & 0 & 0 & 1 & 0 & | & 4 \\ - & - & - & - & - & - & - & | & - \\ -60 & -50 & -40 & 0 & 0 & 0 & 1 & | & 0 \end{bmatrix} \rightarrow$$

$$\begin{bmatrix} u & v & w & r & s & t & M & | & sol \\ 3 & 1 & 0 & 1 & 0 & 0 & 0 & | & 2 \\ \boxed{1} & 0 & .4 & 0 & .2 & 0 & 0 & | & .6 \\ 0 & 1 & 1 & 0 & 0 & 1 & 0 & | & 4 \\ - & - & - & - & - & - & - & | & - \\ -60 & -50 & -40 & 0 & 0 & 0 & 1 & | & 0 \end{bmatrix} \rightarrow \begin{bmatrix} u & v & w & r & s & t & M & | & sol \\ 0 & 1 & -1.2 & 1 & -.6 & 0 & 0 & | & .2 \\ \boxed{1} & 0 & .4 & 0 & .2 & 0 & 0 & | & .6 \\ 0 & 1 & 1 & 0 & 0 & 1 & 0 & | & 4 \\ - & - & - & - & - & - & - & | & - \\ 0 & -50 & -16 & 0 & 12 & 0 & 1 & | & 36 \end{bmatrix} \rightarrow$$

$$\begin{bmatrix} u & v & w & r & s & t & M & | & sol \\ 0 & 1 & -1.2 & 1 & -.6 & 0 & 0 & | & .2 \\ 1 & 0 & \boxed{.4} & 0 & .2 & 0 & 0 & | & .6 \\ 0 & 0 & 2.2 & -1 & .6 & 1 & 0 & | & 3.8 \\ - & - & - & - & - & - & - & | & - \\ 0 & 0 & -76 & 50 & -18 & 0 & 1 & | & 46 \end{bmatrix} \rightarrow \begin{bmatrix} u & v & w & r & s & t & M & | & sol \\ 0 & 1 & -1.2 & 1 & -.6 & 0 & 0 & | & .2 \\ 2.5 & 0 & \boxed{1} & 0 & .5 & 0 & 0 & | & 1.5 \\ 0 & 0 & 2.2 & -1 & .6 & 1 & 0 & | & 3.8 \\ - & - & - & - & - & - & - & | & - \\ 0 & 0 & -76 & 50 & -18 & 0 & 1 & | & 46 \end{bmatrix} \rightarrow$$

$$\begin{bmatrix} u & v & w & r & s & t & M & | & sol \\ 3 & 1 & 0 & 1 & 0 & 0 & 0 & | & 2 \\ 2.5 & 0 & 1 & 0 & .5 & 0 & 0 & | & 1.5 \\ -5.5 & 0 & 0 & -1 & -.5 & 1 & 0 & | & .5 \\ - & - & - & - & - & - & - & | & - \\ 190 & 0 & 0 & 50 & 20 & 0 & 1 & | & 160 \end{bmatrix}$$

The maximum is M = 160 for u = 0, v = 2, and w = 1.5. The original function m has a minimum of m = 160 for x = 50, y = 20, and z = 0. **(Dual problem)**

16.  The problem is to minimize $m = 50x + 40y = [50 \ 40] \begin{bmatrix} x \\ y \\ z \end{bmatrix}$

subject to $10x + 12y \geq 25$

$$15x + 20y \geq 50$$

$$300x + 200y \geq 1000$$

$$x \geq 0, y \geq 0$$

x = the number of servings of food 1, y = the number of servings of food 2.

The constraints are rewritten $\begin{bmatrix} 10 & 12 \\ 15 & 20 \\ 300 & 200 \end{bmatrix} \begin{bmatrix} x \\ y \end{bmatrix} \geq \begin{bmatrix} 25 \\ 50 \\ 1000 \end{bmatrix}, \begin{bmatrix} x \\ y \end{bmatrix} \geq 0.$

This is the dual problem, so solve maximize $M = [25 \ 50 \ 1000] \begin{bmatrix} u \\ v \\ w \end{bmatrix}$

$= 25u + 50v + 1000w$ subject to $\begin{bmatrix} 10 & 15 & 300 \\ 12 & 20 & 200 \end{bmatrix} \begin{bmatrix} u \\ v \\ w \end{bmatrix} \leq \begin{bmatrix} 50 \\ 40 \end{bmatrix}, \begin{bmatrix} u \\ v \\ w \end{bmatrix} \geq 0,$ or

subject to $10u + 15v + 300w \ 50$

$\qquad 12u + 20v + 200w \ 40$

$\qquad u \geq 0, \ v \geq 0, \ w \geq 0$

Introducing the slack variables, the initial tableau becomes

$\begin{bmatrix} u & v & w & s & t & M & sol \\ 10 & 15 & \boxed{300} & 1 & 0 & 0 & 50 \\ 12 & 20 & 200 & 0 & 1 & 0 & 40 \\ -- & -- & -- & -- & - & - & -- \\ -25 & -50 & -1000 & 0 & 0 & 1 & 0 \end{bmatrix} \rightarrow \begin{bmatrix} u & v & w & s & t & M & sol \\ \frac{1}{30} & \frac{1}{20} & \boxed{1} & \frac{1}{300} & 0 & 0 & \frac{1}{6} \\ 12 & 20 & 200 & 0 & 1 & 0 & 40 \\ -- & -- & -- & -- & - & - & -- \\ -25 & -50 & -1000 & 0 & 0 & 1 & 0 \end{bmatrix} \rightarrow$

$\begin{bmatrix} u & v & w & s & t & M & sol \\ \frac{1}{30} & \frac{1}{20} & 1 & \frac{1}{300} & 0 & 0 & \frac{1}{6} \\ \frac{16}{3} & 10 & 0 & -\frac{2}{3} & 1 & 0 & \frac{20}{3} \\ -- & -- & - & -- & - & - & -- \\ \frac{25}{3} & 0 & 0 & \frac{10}{3} & 0 & 1 & \frac{500}{3} \end{bmatrix}$

The maximum is $M = 500/3$ when $u = 0$, $v = 0$, and $w = 1/6$. The minimum cost is $m = 500/3 \approx 166.67$ cents when $x = 10/3$ and $y = 0$. Thus, 3 1/3 servings of food 1 will supply needed nutrients at a minimal cost of about \$1.67 per day. (**Dual problem**)

# Grade Yourself

Circle the numbers of the questions you missed, then fill in the total incorrect for each topic. If you answered more than three questions incorrectly, you need to focus on that topic. (If a topic has less than three questions and you had at least one wrong, we suggest you study that topic also. Read your textbook or a review book, or ask your teacher for help.)

## Subject: Linear Programming

| Topic | Question Numbers | Number Incorrect |
|---|---|---|
| Linear inequalities | 1a, 1b | |
| Geometric solution | 2, 3, 4, 5, 6 | |
| Standard linear programming problem | 7 | |
| Simplex algorithm | 8, 9, 12 | |
| Simplex method and dual problem | 10, 11 | |
| Nonstandard problems | 13, 14 | |
| Dual problem | 15, 16 | |

# Sets and Counting

**4**

---

## Brief Yourself

A set is a collection of elements; $a \in A$ means a is an element of a set A; write $a \notin A$ when a is not in A. A set, A, is contained in a set, B, if all the elements in A are also in B. The containment is proper if A and B are not equal; write $A \subset B$. One way to write a set is to use the listing method: List the elements of the set inside brackets, as the set of digits is $\{0, 1, 2, 3, 4, 5, 6, 7, 8, 9\}$. A second way is to use the descriptive method, as $\{x | x \text{ is a digit}\}$.

Given two sets, A and B, we can form their union, $A \cup B$, the set of elements in A or B (or both), and their intersection, $A \cap B$, the set of elements common to both A and B. A universal set is the domain of discourse. The complement, $A'$, of a set A with respect to a universal set U is the set of elements of U that are not in A. If n(A) = the number of elements of A, then $n(A \cup B) = n(A) + n(B) - n(A \cap B)$.

The number of ways to perform a two-step operation when there are m possible first steps and n possible second steps is mn. This generalizes to multi-step operations.

n!, read "n factorial," is defined to be the product of positive integers from n down to 1, so that $5! = 5 \cdot 4 \cdot 3 \cdot 2 \cdot 1 = 120$. Also, $0! = 1$, by definition.

A permutation of n objects is an arrangement in order of the n objects. There are n! such arrangements. A permutation of k out of n possible objects is an arrangement in order of k of the n objects. There are $n(n-1)(n-2)...(n-k+1) = \dfrac{n!}{(n-k)!}$ of these.

A combination is the number of ways to choose k out of n possible objects, where the order of choice is irrelevant. There are $C(n, k) = \dfrac{n!}{k!(n-k)!}$ ways to do this.

The binomial theorem states that $(x + y)^n =$
$$\binom{n}{0} x^n y^0 + \binom{n}{1} x^{n-1} y^1 + \binom{n}{2} x^{n-2} y^2 + ... + \binom{n}{k} x^{n-k} y^k + ... + \binom{n}{n} x^0 y^n$$

# Test Yourself

1. Write each set using the listing method.

   a. A = {n|n is a positive integer less than 10}

   b. B = {x | x is a vowel}

   c. C = {x | x is a letter coming before j in the alphabet}

   d. D = {x | x is a state of the USA beginning with M}

   e. E = {n | n is a positive multiple of 5}

   f. F = {n | n is a multiple of 7}

   g. G = {x | x is a living unicorn}

2. Write each set using the descriptive method.

   a. A = {A, B, C, D, F}

   b. B = {1, 3, 5, 7, 9}

   c. C = {Alaska, Alabama, Arkansas, Arizona}

   d. D = {2, 4, 6, 8, 10, ...}

   e. E = {0, ± 3, ± 6, ± 9, ± 12, ...}

3. Let A = {1, 2, 3, 4, 5}, B = {1, 3, 5}, and C = {1, 2, 3}, and decide whether each of the following is true or false.

   a. $1 \in A$

   b. $6 \notin A$

   c. $3 \notin B$

   d. $4 \notin C$

   e. $B \subset A$

   f. $A \subset C$

   g. $C \subset B$

   h. $A \subset B$

   i. $\emptyset \subset A$

   j. $\emptyset \in A$

   k. $\emptyset \in \emptyset$

   l. $\emptyset \subset \emptyset$

4. Given U = {1, 2, 3, 4, 5, 6, 7, 8, 9}, A = {1, 2, 3, 4}, B = {1, 3, 5, 9}, and C = {2, 4, 6}, find

   a. $A \cup B$

   b. $A \cap B$

   c. $B \cup C$

   d. $B \cap C$

   e. $A'$

   f. $A' \cup B'$

   g. $B \cap C'$

   h. $(A \cup B)'$

   i. $(A \cup C)'$

   j. $(A' \cap C) \cup B$

   k. $(A \cup B') \cap C'$

   l. $(A \cap C) \cup (A \cap B)$

5. If n(A) = 20, n(B) = 30, and n(A ∩ B) = 8, find n(A ∪ B).

6. If n(A) = 40, n(B) = 30, and n(A ∪ B) = 55, find n(A ∩ B).

7. If there are 500 items and n(A) = 350, find n(A′).

8. Out of 200 items, n(A ∪ B) = 180, n(A ∩ B) = 50, and n(A) = 80. Find n(B′).

9. In the following Venn diagrams, shade the region representing the set.

   a. A′ ∩ B

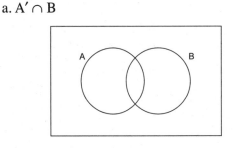

   b. A′ ∩ B′

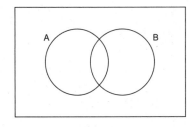

c. $A \cup B'$

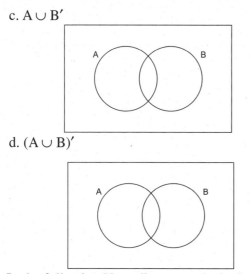

d. $(A \cup B)'$

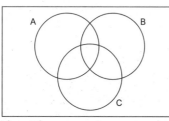

10. In the following Venn diagrams, shade the region representing the set.

a. $(A' \cap B) \cup C$

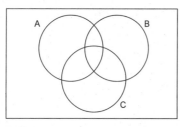

b. $(A \cup B)' \cap C'$

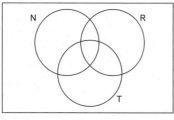

11. Use a Venn diagram to answer the following questions. 45 people read the news, 36 listened to the news on the radio, 35 watched the news on TV, 20 both read the papers and listen to the radio, 14 listened to the radio and watched TV, 13 read the news and watched it on TV, 8 obtained news from all three sources, and 23 obtained news from none of the three sources.

a. How many read the newspaper only?

b. How many obtained news from exactly one of the three sources?

c. How many either read the newspaper or watched on TV?

d. How many people were polled?

12. Use a Venn diagram to answer the following questions. 30% of athletes polled played basketball, 50% played football, and 35% played hockey. 20% played both basketball and football, 16% played both football and hockey, 9% played both basketball and hockey, and 5% played all three.

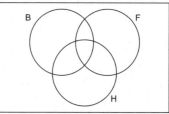

a. What percent played football only?

b. What percent played two or three of the sports?

c. What percent played either basketball or hockey?

d. What percent played none of the three?

13. A fast-food restaurant offers 5 types of sandwiches, 3 types of side orders, 4 types of soft drinks, and 3 types of dessert. How many different meals, each composed of a sandwich, a side order, a soft drink, and a dessert, are possible?

14. Each automobile license plate in a certain state starts with three letters (I and O are not used), followed by three digits. How many license plates are possible?

15. How many of the license plates of problem 10 have no repeated letters or digits?

16. How many of the license plates in problem 10 have at least one repeated letter or digit?

17. There are 6 red marbles and 4 yellow marbles in a box. In how many ways can first a red marble and then a yellow marble be selected?

18. There are 5 new houses in a row on the street, and there are 6 colors of house paint available.

a. In how many ways can the five houses be painted?

b. How many possible ways can they be painted if each house is to have a different color from the one next to it?

c. In how many possible ways can the houses be painted if each house is a different color from each of the others?

19. There are 6 pictures available for hanging in a gallery, but there is only space for 3. How many arrangements of three of the six pictures are possible?

20. A club with 15 members elects a president, a vice president, a secretary, and a treasurer. If no person holds more than one office, in how many ways can the four officers be selected?

21. There are 4 soloists for a certain concert. In how many ways can they be lined up on the stage?

22. There are 6 dignitaries at a ceremony. In how many ways can they be seated in a line across the stage?

23. There are 12 people in a club. In how many ways can a committee of 3 be chosen?

24. There are 80 door prize stubs in a box. In how many ways can 4 be selected to win coffee mugs?

25. There are 50 people in a room. If each person shakes hands once with all the others in the room, how many handshakes are there?

26. There are 8 women and 10 men in a club. In how many ways can a committee

a. of 4 be formed?

b. of 2 men and 2 women be formed?

c. of 4 men be formed?

d. of 4 with only 1 woman be formed?

27. Expand $(x + y)^4$ using the binomial theorem.

28. Expand $(x - 1)^3$ using the binomial theorem.

29. Find the coefficient of $x^3$ in the expansion of $(x + 2)^7$.

30. Find the coefficient of $x^4 y^3$ in the expansion of $(2x - 3y)^7$.

# ✔ Check Yourself

1.  a. A = {1, 2, 3, 4, 5, 6, 7, 8, 9}

b. B = {a, e, i, o, u}, or {a, e, i, o, u, y}

c. C = {a, b, c, d, e, f, g, h, i}

d. D = {Maine, Maryland, Massachusetts, Michigan, Minnesota, Mississippi, Missouri, Montana}

e. E = {5, 10, 15, 20, 25, 30 ...}

f. F = {0, ± 7, ± 14, ± 21 ...}

g. G = ∅ **(Sets)**

2.  a. A = {x | x is a letter grade}

b. B = {n | n is an odd digit}

c. C = {x | x is a state of the USA beginning with an A}

d. D = {n | n is an even positive integer}

e. E = {n | n is a multiple of 3} **(Sets)**

3.   a. $1 \in A$ True since 1 is on the list.

   b. $6 \notin A$ True since 6 is not on the list.

   c. $3 \notin B$ False since 3 is on the list for B.

   d. $4 \notin C$ True since 4 is not on the list for C.

   e. $B \subset A$ True since each element of B is also in A.

   f. $A \subset C$ False since 4 is in A but 4 is not in C.

   g. $C \subset B$ False since 2 is in C but 2 is not in B.

   h. $A \subset B$ False since 2 is in A but 2 is not in B.

   i. $\varnothing \subset A$ True since $\varnothing$ is a subset of every set.

   j. $\varnothing \in A$ False since $\varnothing$ is not on the list for A.

   k. $\varnothing \in \varnothing$ False since $\varnothing$ contains no elements.

   l. $\varnothing \subset \varnothing$ True since $\varnothing$ is a subset of every set.   **(Sets)**

4.   a. $A \cup B = \{1, 2, 3, 4, 5, 9\}$

   b. $A \cap B = \{1, 3\}$

   c. $B \cup C = \{1, 2, 3, 4, 5, 6, 9\}$

   d. $B \cap C = \varnothing$

   e. $A' = \{5, 6, 7, 8, 9\}$

   f. $A' \cup B = \{2, 4, 5, 6, 7, 8, 9\}$

   g. $B \cap C' = \{1, 3, 5, 9\}$

   h. $(A \cup B)' = \{6, 7, 8\}$

   i. $(A \cup C)' = \{5, 7, 8, 9\}$

   j. $(A' \cap C) \cup B = \{6\} \cup \{1, 3, 5, 9\} = \{1, 3, 5, 6, 9\}$

   k. $(A \cup B') \cap C' = \{1, 2, 3, 4, 6, 7, 8\} \cap \{1, 3, 5, 7, 8, 9\} = \{1, 3, 7, 8\}$

   l. $(A \cap C) \cup (A \cap B) = \{2, 4\} \cup \{1, 3\} = \{1, 2, 3, 4\}$. **(Intersection, union, and complement)**

5.   $n(A \cup B) = n(A) + n(B) - n(A \cup B) = 20 + 30 - 8 = 42$. **(Intersection, union, and complement)**

6.   $n(A \cup B) = 55 = n(A) + n(B) - n(A \cap B) = 40 + 30 - n(A \cap B)$, so that $55 = 70 - n(A \cap B)$. Thus, $n(A \cap B) = 70 - 55 = 15$. **(Intersection, union, and complement)**

7.   $n(A') = $ total number $- n(A) = 500 - 350 = 150$. **(Intersection, union, and complement)**

8.   $n(A \cup B) = 180 = n(A) + n(B) - n(A \cap B) = 80 + n(B) - 50$. Thus, $180 = 30 + n(B)$, so $n(B) = 150$. $n(B') = 200 - 150 = 50$. **(Intersection, union, and complement)**

9.  a. A′ ∩ B

b. A′ ∩ B′

c. A ∪ B′

d. (A ∪ B)′

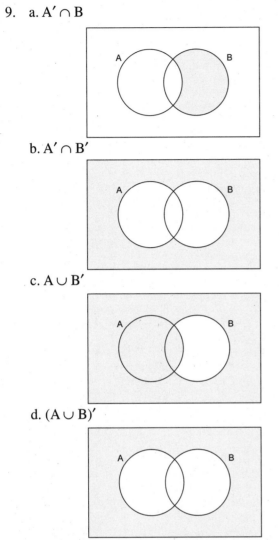

**(Intersection, union, and complement)**

10. a. (A′ ∩ B) ∪ C

b. (A ∪ B)′ ∩ C′

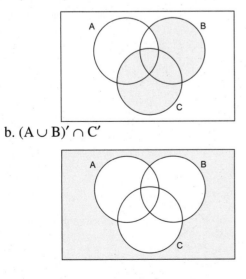

**(Intersection, union, and complement)**

11.

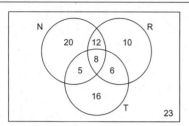

a. newspaper only 20

b. exactly one 20 + 10 + 16 = 46

c. read the newspaper or watched TV 45 + 16 + 6 = 67

d. people polled 45 + 10 + 6 + 16 + 23 = 100

**(Intersection, union, and complement)**

12.

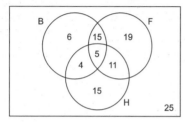

a. football only 19%

b. two or three 15 + 5 + 4 + 11 = 35%

c. either basketball or hockey 30 + 11 + 15 = 56%

d. played none 25%   **(Intersection, union, and complement)**

13.  $5 \times 3 \times 4 \times 3 = 180$ **(Multiplication principle)**

14.  $24 \times 24 \times 24 \times 10 \times 10 \times 10 = 13,824,000$ **(Multiplication principle)**

15.  $24 \times 23 \times 22 \times 10 \times 9 \times 8 = 874,680$ **(Multiplication principle)**

16.  $13,824,000 - 874,680 = 12,949,320$ **(Multiplication principle)**

17.  $6 \times 4 = 24$ **(Multiplication principle)**

18.  a. $6 \times 6 \times 6 \times 6 \times 6 = 7776$ **(Multiplication principle)**

b. $6 \times 5 \times 5 \times 5 \times 5 = 3750$ since after the first house, there are only 5 choices of color.  **(Multiplication principle)**

c. $6 \times 5 \times 4 \times 3 \times 2 = 720$, or $\dfrac{6!}{1!} = 720$ **(Permutations)**

19.  $6 \times 5 \times 4 = 120$, or $\dfrac{6!}{3!} = 120$ **(Permutations)**

20.  $\dfrac{15!}{11!} = 15 \times 14 \times 13 \times 12 = 32760$ **(Permutations)**

21.  $4! = 4 \times 3 \times 2 \times 1 = 24$ **(Permutations)**

22. $6! = 6 \times 5 \times 4 \times 3 \times 2 \times 1 = 720$ **(Permutations)**

23. $C(12, 3) = \dbinom{12}{3} = \dfrac{12!}{3!9!} = \dfrac{12 \cdot 11 \cdot 10}{3 \cdot 2 \cdot 1} = 2 \times 11 \times 10 = 220$ **(Combinations)**

24. $C(80, 4) = \dbinom{80}{4} = \dfrac{80!}{4!76!} = \dfrac{80 \cdot 79 \cdot 78 \cdot 77}{4 \cdot 3 \cdot 2 \cdot 1} = 10 \times 79 \times 26 \times 77 = 1{,}581{,}580.$ **(Combinations)**

25. $C(50, 2) = \dbinom{50}{2} = \dfrac{50!}{2!48!} = \dfrac{50 \cdot 49}{2 \cdot 1} = 25 \times 49 = 1225$ **(Combinations)**

26. a. $C(18, 4) = \dbinom{18}{4} = \dfrac{18!}{4!14!} = \dfrac{18 \cdot 17 \cdot 1 \cdot 15}{4 \cdot 3 \cdot 2 \cdot 1} = 3 \times 17 \times 4 \times 15 = 3060$ **(Combinations)**

    b. $C(8, 2) \times C(10, 2) = \dbinom{8}{2} \times \dbinom{10}{2} = \dfrac{8!}{2!6!} \times \dfrac{10!}{2!8!} = \dfrac{8 \cdot 7}{2 \cdot 1} \times \dfrac{10 \cdot 9}{2 \cdot 1} = 28 \times 45 = 1260$ **(Combinations and**

    **multiplication principle)**

    c. $C(10, 4) = \dbinom{10}{4} = \dfrac{10!}{4!6!} = \dfrac{10 \cdot 9 \cdot 8 \cdot 7}{4 \cdot 3 \cdot 2 \cdot 1} = 10 \times 3 \times 1 \times 7 = 210$ **(Combinations)**

    d. $C(8, 1) \times C(10, 3) = \dbinom{8}{1} \times \dbinom{10}{3} = \dfrac{8!}{1!7!} \times \dfrac{10!}{3!7!} = \dfrac{8}{1} \times \dfrac{10 \cdot 9 \cdot 8}{3 \cdot 2 \cdot 1} = 8 \times 10 \times 3 \times 4 = 960$

    **(Combinations and multiplication principle)**

27. $(x + y)^4 = \dbinom{4}{0}x^4 y^0 + \dbinom{4}{1}x^3 y^1 + \dbinom{4}{2}x^2 y^2 + \dbinom{4}{3}x^1 y^3 + \dbinom{4}{4}x^0 y^4$

    $= \dfrac{4!}{0!4!} x^4 + \dfrac{4!}{1!3!} x^3 y + \dfrac{4!}{2!2!} x^2 y^2 + \dfrac{4!}{3!1!} xy^3 + \dfrac{4!}{4!0!} y^4$

    $= x^4 + 4x^3 y + \dfrac{4 \cdot 3}{2 \cdot 1} x^2 y^2 + 4xy^3 + y^4$

    $= x^4 + 4x^3 y + 6x^2 y^2 + 4xy^3 + y^4$ **(Binomial theorem)**

28. $(x - 1)^3 = (x + (-1))^3$

    $= \dbinom{3}{0}x^3 (-1)^0 + \dbinom{3}{1}x^2 (-1)^1 + \dbinom{3}{2}x^1 (-1)^2 + \dbinom{3}{3}x^0 (-1)^3$

    $= \dfrac{3!}{0!3!} x^3 + \dfrac{3!}{1!2!} x^2 (-1) + \dfrac{3!}{2!1!} x + \dfrac{3!}{3!0!} (-1)$

    $= x^3 - 3x^2 + 3x - 1$ **(Binomial theorem)**

29. The term is $\dbinom{7}{4}x^3 (2)^4 = \dfrac{7!}{4!3!} x^3 (16) = \dfrac{7 \cdot 6 \cdot 5}{3 \cdot 2 \cdot 1}(16)x^3 = (35)(16)x^3 = 560x$, so the coefficient is 560.

    **(Binomial theorem)**

30. $(2x - 3y)^7 = ((2x) + (-3y))^7$. The term is $\dbinom{7}{3}(2x)^4 (-3y)^3 = \dfrac{7!}{3!4!} (16x^4)(-27y^3) = \dfrac{7 \cdot 6 \cdot 5}{3 \cdot 2 \cdot 1}(-432)x^4 y^3 =$

    $(35)(-432)x^4 y^3 = -15120x^4 y^3$, so the coefficient is $-15120$. **(Binomial theorem)**

# Grade Yourself

Circle the numbers of the questions you missed, then fill in the total incorrect for each topic. If you answered more than three questions incorrectly, you need to focus on that topic. (If a topic has less than three questions and you had at least one wrong, we suggest you study that topic also. Read your textbook or a review book, or ask your teacher for help.)

## Subject: Sets and Counting

| Topic | Question Numbers | Number Incorrect |
|---|---|---|
| Sets | 1, 2, 3 | |
| Intersection, union, and complement | 4, 5, 6, 7, 8, 9, 10, 11, 12 | |
| Multiplication principle | 13, 14, 15, 16, 17, 18a, 18b | |
| Permutations | 18c, 19, 20, 21, 22 | |
| Combinations | 23, 24, 25, 26 | |
| Binomial theorem | 27, 28, 29, 30 | |

# *Probability*

## 5

## Brief Yourself

The set of all possible outcomes of an experiment is the sample space. If the outcomes are equiprobable and if there are n outcomes in the sample space, then the probability of each one is $\frac{1}{n}$. An event is a subset of the sample space. If an event contains k outcomes, and the outcomes are equiprobable, then the probability of E, $p(E) = \frac{k}{n}$. In any case, each probability, p, must satisfy $0 \leq p \leq 1$, and the sum of the probabilities of all the outcomes must be 1. Generally, the probability of an event, E, is the sum of the probabilities of the outcomes in E.

If E and F are events, then $p(E \cup F) = p(E) + p(F) - p(E \cap F)$. If E and F are mutually exclusive, then they can not both occur. This means that $p(E \cap F) = 0$, so that for mutually exclusive events, $p(E \cup F) = p(E) + p(F)$. Also, $p(E') = 1 - p(E)$, where $E'$ is the complement of E.

The probability of an event, F, given that another event, E, has occurred, is $p(F|E) = n\frac{(E \cap F)}{n(E)} = p\frac{(E \cap F)}{p(E)}$.

Two events, E and F, are independent if the occurrence of one does not change the probability that the other will occur. If E and F are independent, then $p(E \cap F) = p(E)p(F)$. Conversely, to show that E and F are independent, check to see that $p(E \cap F) = p(E)p(F)$.

Bayes' Theorem states that if $E_1, E_2 \ldots E_k$ are mutually exclusive events with $E_1 \cup E_2 \cup \ldots \cup E_k = S$, and if F is any event in S, then

$$p(E|F) = \frac{p(F|E_1)\,p(E_1)}{p(F|E_1)\,p(E_2) + p(F|E_1)\,p(E_2) + \ldots + p(F|E_k)\,p(E_k)}$$

Bernoulli probabilities are computed using the formula $p(x \text{ successes in } n \text{ trials}) = C(n, x)p^x\,q^{n-x}$, where $p = p(\text{success})$ in one trial, $p + q = 1$, n = the number of trials. Note that trials must be independent.

# Test Yourself

1.  For each of the following experiments, write down the sample space.

    a. A vowel is selected.

    b. A coin is tossed twice, heads or tails is recorded.

    c. A single die is tossed.

    d. A pair of dice is tossed, and the sum is recorded.

    e. Of three marbles in an urn, one red, one green, and one blue, two are selected and the colors recorded.

2.  Decide whether each of the following is a valid assignment of probabilities, if the sample space is {a, b, c, d}. Explain your answers.

    a. p(a) =.1, p(b) =.2, p(c) =.3, p(d) =.4

    b. p(a) = 0, p(b) = -.4, p(c) =.7, p(d) =.7

    c. $p(a) = \frac{1}{12}$, $p(b) = \frac{1}{6}$, $p(c) = \frac{1}{4}$, $p(d) = \frac{1}{2}$

    d. $p(a) = \frac{1}{5}$, $p(b) = \frac{2}{5}$, $p(c) = \frac{2}{5}$, $p(d) = \frac{1}{5}$

    e. p(a) = 1, p(b) = 0, p(c) = 0, p(d) = 0

    f. p(a) =.45, p(b) =.24, p(c) =.11, p(d) =.18

3.  Use a tree to find all the outcomes in the sample space.

    a. First a die is tossed, and then a coin is tossed.

    b. A coin is tossed three times.

4.  Alice was born in September. What is the probability that she was born

    a. on the first day of the month?

    b. during the last week?

5.  There are 5 white marbles, 4 green marbles, and 3 yellow marbles in a box. If 1 is selected, what is the probability that it is yellow? Green or white? Red?

6.  There are 20 freshmen, 14 sophomores, 16 juniors, and 12 seniors in a club. If a member is selected at random, what is the probability that he/she is a freshman? A junior or a senior?

7.  A coin is tossed three times. What is the probability that heads comes up exactly once? At least once?

8.  There are 6 red marbles and 4 green marbles in a box. Two are selected (without replacement). Find the probability that the first is red and the second is green and the probability that both are green.

9.  A red die and a green die are tossed, and the number on each is recorded (red first, green second). Write down the outcomes in the event E is "red came up 2" and the event F is "the sum is 5."

10. Find the probabilities of the events E and F of problem 9.

11. A child has probability .45 of owning a Gameboy, probability .53 of owning Sega, and probability .13 of owning both.

    a. What is the probability that the child owns either a Gameboy or a Sega?

    b. What is the probability that he/she owns neither one?

12. The probability that a toy train has a wheel defect is .09, a whistle defect .05, and both .02.

    a. Find the probability that the train has either a wheel defect or a whistle defect.

    b. Find the probability that a train with a wheel defect also has a whistle defect.

13. Which of the following are mutually exclusive events?

    a. E is "an integer is a multiple of 5," and F is "an integer is a multiple of 11."

    b. E is "a person is born in February," and F is "a person is born on the 30th of the month."

    c. E is "a student is taking English," and F is "a student is taking math."

    d. E is "the sum on a pair of dice is 9," F is "doubles came up."

14. Suppose that E and F are events, with p(E) =.4 and p(F) =.5.

    a. Find p(E ∪ F) if p(E ∩ F) =.3.

b. Find p(E ∪ F) if E and F are mutually exclusive.

c. Find p(E ∪ F) if E and F are independent.

d. Find p(E | F) if E and F are independent.

e. If p(E ∩ F) = .3, find p(F | E).

15. There are 25 students in the first-period physics class, and of these 15 received an A on the first quiz. There are 30 students in the second-period physics class, and 25 received an A on the first quiz.

a. What is the probability that a student chosen at random from the first period got an A?

b. What is the probability that a student chosen at random from the second period got an A?

c. What is the probability that a student chosen at random from one of the two classes got an A?

d. What is the probability that a student chosen at random from those that received an A is in the first-period class?

16. There is a pile of 60 sweatshirts on a table, all on sale.

| | Plain Red | Plain Blue | Decorated Blue |
|---|---|---|---|
| Small | 8 | 5 | 3 |
| Medium | 8 | 2 | 10 |
| Large | 2 | 3 | 5 |
| Extra Large | 4 | 5 | 5 |

If one sweatshirt is chosen at random, what is

a. the probability that it is red?

b. the probability that it is plain?

c. the probability that it is plain and extra large?

d. the probability that it is small, given that it is red?

e. the probability that it is large and decorated, given that it is blue and not extra large?

f. If E is the event "the sweatshirt is medium" and F is the event "the sweatshirt is blue," are E and F independent, and how can you tell?

17. The table below shows the percentages of various types of books in a library.

| | Fiction | Nonfiction | Biography |
|---|---|---|---|
| pre-1980 | 12% | 15% | 5% |
| 1980-1989 | 7% | 12% | 9% |
| 1990 or after | 12% | 18% | 10% |

If a book is chosen at random from this library, find the probability that

a. the book is either fiction or biography.

b. the book was published in the 1980s.

c. the book is nonfiction given that it was published in the 1990s.

d. the book is a biography published before 1980.

e. Are the events "the book is a 1990s publication" and "the book is nonfiction" independent? How can you tell?

f. Are the events "the book was published before 1990" and "the book is a biography" independent? How can you tell?

18. The probability of rain was 40%. If it rained, the Redskins had a 30% chance of winning; if it did not rain, they had a 55% chance of winning. Given that the Redskins won, what is the probability that it rained?

19. A rare disease affects 5% of the population. A test for this disease is positive 4% of the time when no disease is present and negative 2% of the time when the disease is present. Given that the test is positive, what is the probability that the disease is present?

20. A calculator company has three plants. The New York plant manufactures 40% of the calculators, the Louisiana plant 35%, and the Iowa plant 25%. Five percent of the calculators manufactured in New York are defective, 3% of those manufactured in Louisiana are defective, and 2% of those manufactured in Iowa are defective. A defective calculator is discovered. What is the probability that it came from the Iowa plant?

21. The probability that a properly planted maple sapling lives for the next 5 years is 80%. If 12 saplings are planted, what is the probability that 10 live for the next 5 years?

22. A certain treatment is successful in 75% of the cases of ulcers. If 7 patients are treated,

    a. What is the probability that 5 are treated successfully?

    b. What is the probability that at least 5 are successfully treated?

    c. What is the probability that at most 2 are successfully treated?

    d. What is the probability that at least 3 are successfully treated?

# ✓ Check Yourself

1. a. {a, e, i, o, u} or {a, e, i, o, u, y}

   b. {HH, HT, TH, TT}, where H stands for heads and T for tails.

   c. {1, 2, 3, 4, 5, 6}

   d. {2, 3, 4, 5, 6, 7, 8, 9, 10, 11, 12}

   e. {RG, RB, GR, GB, BR, BG}, where R stands for red, G for green, and B for blue. **(Basic ideas)**

2. a. Yes; all probabilities are between 0 and 1, and they sum up to 1.

   b. No; p(b) is not between 0 and 1.

   c. Yes; all probabilities are between 0 and 1, and they sum up to 1.

   d. No; the probabilities do not sum up to 1.

   e. Yes; all probabilities are between 0 and 1, and they sum up to 1.

   f. No; the probabilities do not sum up to 1. **(Basic ideas)**

3. a.

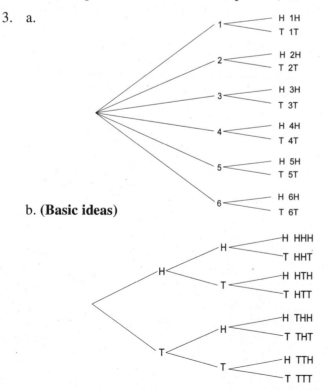

   b. **(Basic ideas)**

4.  a. p(born on the first) = $\frac{1}{30}$

    b. p(born during the first week) = $\frac{7}{30}$. **(Basic ideas)**

5.  p(yellow) = $\frac{3}{12}$ = $\frac{1}{4}$ , p(green or white) = $\frac{9}{12}$ = $\frac{3}{4}$ , p(red) = 0. **(Basic ideas)**

6.  p(freshman) = $\frac{20}{62}$ = $\frac{10}{31}$ , p(junior or senior) = $\frac{28}{62}$ = $\frac{14}{31}$. **(Basic ideas)**

7.  p(heads once) = p(HTT or THT or TTH) = $\frac{3}{8}$ , since there are 8 possible outcomes. (See problem 3b.)

    p(at least once) = 1 − p(no times) = 1 − p(TTT) = 1 − $\frac{1}{8}$ = $\frac{7}{8}$. **(Basic ideas and complementary events)**

8.  p(RG) = $\frac{6\times4}{10\times9}$ = $\frac{24}{90}$ = $\frac{4}{15}$, or p(RG) = $\frac{C(6,\,1)\times C(4,\,1)}{C(10,\,2)}$ = $\frac{4}{15}$, p(GG) = $\frac{4\times3}{10\times9}$ = $\frac{2}{15}$, or p(GG) = $\frac{C(4,2)}{C(10,2)}$

    = $\frac{2}{15}$. **(Basic ideas)**

9.  E is "red came up 2" =

    {(2, 1), (2, 2), (2, 3), (2, 4), (2, 5), (2, 6)}

    F is "the sum is 5" = {(1, 4), (2, 3), (3, 2), (4, 1)}. **(Basic ideas)**

10. p(E) = $\frac{6}{36}$ = $\frac{1}{6}$ , since there are 36 possible outcomes for the two dice (6 × 6 = 36, according to the

    multiplication principle). p(F) = $\frac{4}{36}$ = $\frac{1}{9}$. **(Basic ideas)**

11. a. p(Gameboy ∪ Sega)

    = p(Gameboy) + p(Sega) − p(Gameboy ∩ Sega) =.45 +.53 -.13 =.85. **(Addition principle)**

    b. p(neither) = 1 − p(Gameboy or Sega) = 1 − .85 =.15. **(Complementary events)**

12. a. p(wheel or whistle defect)

    = p(wheel defect) + p(whistle defect) − p(both)

    =.09 +.05 − .02 =.12. **(Addition principle)**

    b. p(whistle|wheel) = $\frac{p(both)}{p(wheel)}$ = $\frac{.02}{.09}$ = $\frac{2}{9}$. **(Conditional probability)**

13. Which of the following are mutually exclusive events?

    a. Not mutually exclusive; 55 is a multiple of both.

    b. Mutually exclusive; a person cannot be born on the 30th of February.

    c. Not mutually exclusive; a person can take both English and math.

    d. Mutually exclusive; if doubles comes up, the sum cannot be 9. **(Mutually exclusive events)**

14. a. $p(E \cup F) = p(E) + p(F) - p(E \cap F) = .4 + .5 - .3 = .6.$ (**Addition principle**)

    b. $p(E \cup F) = p(E) + p(F) - p(E \cap F) = .4 + .5 - 0 = .9.$ (**Addition principle**)

    c. $p(E \cup F) = p(E) + p(F) - p(E \cap F) = .4 + .5 - (.4)(.5) = .7.$ (**Independent events**)

    d. $p(E \mid F) = p(E) = .4$, if E and F are independent. (**Independent events**)

    e. $p(F \mid E) = \dfrac{p(E \circ F)}{p(E)} = \dfrac{.3}{.4} = .75.$ (**Conditional probability**)

15. a. p(student in first period got an A) $= \dfrac{15}{25} = \dfrac{3}{5}$

    b. p(student in second period got an A) $= \dfrac{25}{30} = \dfrac{5}{6}$

    c. p(student got an A) $= \dfrac{15+25}{25+30} = \dfrac{40}{55} = \dfrac{8}{11}$. (**Basic Ideas**)

    d. p(student in first $\mid$ student got an A) $= \dfrac{p(\text{both})}{p(A)} = \dfrac{15}{40} = \dfrac{3}{8}$. (**Conditional probability**)

16. a. p(red) $= \dfrac{8+8+2+4}{60} = \dfrac{22}{60} = \dfrac{11}{30}$

    b. p(plain) $= \dfrac{8+8+2+4+5+2+3+5}{60} = \dfrac{37}{60}$

    c. p(plain and extra large) $= \dfrac{4+5}{60} = \dfrac{9}{60} = \dfrac{3}{20}$. (**Basic ideas**)

    d. p(small $\mid$ red) $= \dfrac{n(\text{small and red})}{n(\text{red})} = \dfrac{8}{8+8+2+4} = \dfrac{8}{22} = \dfrac{4}{11}$ (**Conditional probability**)

    e. p(large and decorated $\mid$ blue and not extra large) $= \dfrac{n(\text{large, decorated, blue, not XL})}{n(\text{blue and not XL})} = \dfrac{5}{5+2+3+3+10+5} = \dfrac{5}{28}$. (**Conditional probability**)

    f. $p(E) = \dfrac{8+2+10}{60} = \dfrac{1}{3}$, $p(F) = \dfrac{5+2+3+5+3+10+5+5}{60} = \dfrac{38}{60} = \dfrac{19}{30}$, p(E and F) $= \dfrac{2+10}{60} = \dfrac{1}{5}$.

    For independence, check to see whether $p(E \cap F) = p(E)p(F)$. In this case, $\dfrac{1}{5} \neq \dfrac{1}{3} \times \dfrac{19}{30}$, so the events are not independent. (**Independent events**)

17. a. p(fiction or biography) $= .12 + .07 + .12 + .05 + .09 + .10 = .55$

    b. p(published in the 1980's) $= .07 + .12 + .09 = .28.$ (**Addition principle**)

    c. p(nonfiction $\mid$ published in the 1990's) $= \dfrac{.18}{.12 + .18 + .10} = \dfrac{.18}{.40} = .45.$ (**Conditional probability**)

    d. p(biography published before 1980) $= .05.$ (**Basic ideas**)

    e. p(the book is a 1990's publication) $= .12 + .18 + .10 = .40$, p(the book is nonfiction) $= .15 + .12 + .18 = .45$, p(both) $= .18 = .4 \times .45 = $ p(1990's) $\times$ p(nonfiction), so the events are independent.

f. p(the book was published before 1990) =.12 +.07 +.15 +.12 +.05 +.09 =.60

p(the book is a biography) =.05 +.09 +.10 =.24

p(both) =.05 +.09 =.14 $\neq$ .6 $\times$ .24, so the events are not independent. (**Independent events**)

18. p(rained | won) =

$$\frac{p(\text{won} \mid \text{rained}) \times p(\text{rained})}{p(\text{won} \mid \text{rained}) \times p(\text{rained}) + p(\text{won} \mid \text{no rain}) \times p(\text{no rain})}$$

$$\frac{.3 \times .4}{.3 \times .4 + .55 \times .6} = \frac{.12}{.12 + .33} = \frac{.12}{.45} = =.2667 \quad (\textbf{Bayes' theorem})$$

19. p(disease | test positive) =

$$\frac{p(\text{test pos} \mid \text{dis}) \times p(\text{dis})}{p(\text{test pos} \mid \text{dis}) \times p(\text{dis}) + p(\text{test pos} \mid \text{no dis}) \times p(\text{no dis})} = \frac{.98 \times .05}{.98 \times .05 + .04 \times .95} = \frac{.049}{.049 + .038} = \frac{.049}{.087} = .5632$$

(**Bayes' theorem**)

20. p(Iowa | defective) =

$$\frac{p(\text{def} \mid \text{Iowa}) \times p(\text{Iowa})}{p(\text{def} \mid \text{NY}) \times p(\text{NY}) + p(\text{def} \mid \text{LA}) \times p(\text{LA}) + p(\text{def} \mid \text{Iowa}) \times p(\text{Iowa})} = \frac{.02 \times .25}{.05 \times .40 + .03 \times .35 + .02 \times .25} =$$

$$\frac{.005}{.02 + .0105 + .005} = \frac{.005}{.0355} = .1408 \quad (\textbf{Bayes' theorem})$$

21. p(10 successes in 12 trials) = C(12, 10)$.8^{10} .2^2 = \frac{12!}{10!2!}(.1073741824)(.04) = \frac{12 \times 11}{2}(.0042949673) = .2835$

(**Bernoulli trials**)

22. A certain treatment is successful in 75% of the cases of ulcers. If 7 patients are treated:

a. p(5 successes in 7 trials) = C(7, 5)$.75^5 .25^2 = \frac{7!}{5!2!}(.2373046875)(.0625) = \frac{7 \times 6}{2}(.014831543) = .3115$

b. p(at least 5 successes) =

p(5 successes) + p(6 successes) + p(7 successes) =

$.3115 +$ C(7, 6)$.75^6 .25^1 +$ C(7, 7)$.75^7 .25^0 = .3115 + \frac{7!}{6!1!}(.1779785156)(.25) + \frac{7!}{7!0!}(.1334838867) =$

$.3115 + .3115 + .1335 = .7565$

c. p(at most 2 successes) =

p(0 successes) + p(1 success) + p(2 successes) =

C(7, 0)$.75^0 .25^7 +$ C(7, 1)$.75^1 .25^6 +$ C(7, 2)$.75^2 .25^5 = \frac{7!}{0!7!}(.000061035) + \frac{7!}{1!6!}(.75)(.000244141) +$

$\frac{7!}{2!5!}(.5625)(.00097656) = .0129$

d. p(at least 3 successes) = 1 - p(at most 2 successes) = 1 - .0129 = .9871 (**Bernoulli trials**)

# Grade Yourself

Circle the numbers of the questions you missed, then fill in the total incorrect for each topic. If you answered more than three questions incorrectly, you need to focus on that topic. (If a topic has less than three questions and you had at least one wrong, we suggest you study that topic also. Read your textbook or a review book, or ask your teacher for help.)

## Subject: Probability

| Topic | Question Numbers | Number Incorrect |
|---|---|---|
| Basic ideas | 1, 2, 3, 4, 5, 6, 8, 9, 10, 15a-c, 16a-c, 17d, | |
| Basic ideas and complementary events | 7 | |
| Complementary events | 11b | |
| Addition principle | 11a, 12a, 14a-b, 17a-b, | |
| Conditional probability | 12b, 14e, 15d, 16d-e, 17c | |
| Mutually exclusive events | 13 | |
| Independent events | 14c-d, 16f, 17e-f, | |
| Bayes' theorem | 18, 19, 20 | |
| Bernoulli trials | 21, 22 | |

# *Statistics*

**6**

---

## Brief Yourself

Numerical data is often organized in frequency tables and displayed using a histogram. When the numerical data is not all the possible data, it is called a sample.

Common measures of central tendency are the mean, which is the ordinary average; the median, which is the middle value; and the mode, which is the most frequent value. In finding the median, always start by putting the numbers in order. Then the median is the middle number (odd number of data) or the average of the two middle numbers (even number of data).

Common measures of dispersion are the range and the standard deviation. The range is the highest data value minus the lowest data value. To find the variance (1) square each of the data values and add them up; (2) find the mean and square it; (3) subtract the number of data times the square of the mean (2) from the sum of the squares (1); and divide by the number of data values minus 1. The standard deviation is the square root of the variance. Thus:

$$s = \sqrt{\frac{\Sigma(x_i - \overline{x})^2}{n-1}} = \sqrt{\frac{\Sigma x_i^2 - n\overline{x}^2}{n-1}}$$

A probability distribution for a discrete random variable is often given in a table. A list of possible data values is paired with the probability of obtaining that value.

The expected value of a random variable is its mean. To find this, multiply each value by its probability and add these resulting numbers. To find the variance of the random variable, (1) square all the values; (2) multiply the squares by the probabilities of the values and add up the products; and (3) subtract the square of the mean. The standard deviation is the square root of the variance.

Continuous random variables have distributions given by functions, and the probability of obtaining a value in an interval is expressed as an area. The normal distribution is continuous, and a table of values is given for areas under its curve.

The binomial distribution is a discrete distribution, and the probability of x successes in n trials is $C(n, x)p^x q^{n-x}$, a formula that was used in the probability chapter. Binomial probabilities can be approximated using the normal distribution. The mean is np, and the standard deviation is $\sqrt{npq}$.

# Test Yourself

1. Given the following data: 1, 3, 4, 5, 1, 3, 1, 3, 5

   a. Organize the data in a frequency table.

   b. Draw its histogram.

   c. Find the mean.

   d. Find the median.

   e. Find the mode.

   f. Find the range.

   g. Find the variance.

   h. Find the standard deviation.

2. Given the following data:

| x | 5 | 4 | 3 | 2 | 1 | 0 |
|---|---|---|---|---|---|---|
| f | 10 | 12 | 8 | 4 | 8 | 8 |

   a. Draw its histogram.

   b. Find the mean.

   c. Find the median.

   d. Find the mode.

   e. Find the range.

   f. Find the variance.

   g. Find the standard deviation.

3. Given the following histogram:

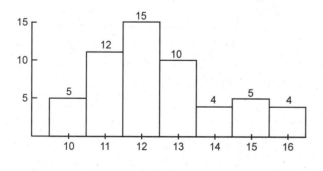

*Fig. 6.1*

   a. Find the frequency table.

   b. Find the mean.

   c. Find the median.

   d. Find the mode.

   e. Find the range.

   f. Find the variance.

   g. Find the standard deviation.

4. Organize the following data into five classes:

   19, 16, 18, 17, 10, 10, 11, 10, 14, 14,
   13, 10, 10, 19, 18, 17, 16, 19, 12, 14

   a. Find the frequency table.

   b. Draw its histogram.
      Use the frequency table to:

   c. Estimate the mean.

   d. Find the class containing the median.

   e. Find the modal class.

   f. Estimate the range.

   g. Estimate the variance.

   h. Estimate the standard deviation.

5. Given the following classed data:

| x | 200–299 | 300–399 | 400–499 | 500–599 | 600–699 | 700–799 |
|---|---|---|---|---|---|---|
| d | 200 | 250 | 300 | 250 | 200 | 100 |

   Interpret the first class to run from 200 to 300, including the 200 and excluding the 300, so that middle number of class is 250. Interpret the other classes similarly.

   a. Draw its histogram.
      Using the frequency table:

   b. Estimate the mean.

   c. Find the class containing the median.

   d. Find the modal class.

   e. Estimate the range.

   f. Estimate the variance.

   g. Estimate the standard deviation.

6. Given the following histogram:

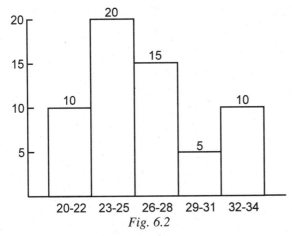

Fig. 6.2

a. Find the frequency table.

b. Estimate the mean.

c. Estimate the median.

d. Find the modal class.

e. Estimate the range.

f. Estimate the variance.

g. Estimate the standard deviation.

7. 3 red balls and 2 green balls are in a box. Two balls are drawn without replacement, and X = the number of green balls drawn.

a. What are the possible values of X?

b. Find the probability distribution.

c. Find the expected value.

d. Find the variance.

e. Find the standard deviation.

8. Three coins are tossed, and X = the number of heads.

a. What are the possible values of X?

b. Find the probability distribution.

c. Find the expected value.

d. Find the variance.

e. Find the standard deviation.

9. A lawyer estimates that it takes her 1 hour to draw up a simple will, 3 hours to draw up a moderately complex will, and 10 hours to draw up a complicated will. If 70% of the wills she draws up are simple, 25% moderately complex, and 10% complicated, about how much time should she expect to spend drawing up 80 wills?

10. You play a game in which you spin a spinner whose arrow can point to any one of three colors with equal probability. If the spinner arrow points to red, you win $3; if it points to green, you win $1; but if it points to yellow, you lose $2.

a. Is the game fair? How can you tell?

b. If the game is not fair, does it favor you or your opponent? Explain.

11. Given the following probability distribution:

| x | 0 | 1 | 2 | 3 | 4 | 5 |
|---|---|---|---|---|---|---|
| p |  | .3 | .1 | .1 | .05 | .05 |

a. Find p(0).

b. Find the mean.

c. Find the standard deviation.

12. The following table gives the number of TV sets per family for 80 families.

| x | 0 | 1 | 2 | 3 | 4 | 5 |
|---|---|---|---|---|---|---|
| f | 5 | 30 | 20 | 15 | 7 | 3 |
| p |  |  |  |  |  |  |

a. Find the probability distribution.

b. Draw its histogram.

13. Given the following probability histogram representing the number of strokes golfers took to complete a certain hole:

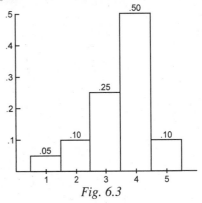

Fig. 6.3

a. What is the probability that a golfer took more than 2 strokes?

b. What is the most common number of strokes taken to complete the hole?

c. What is the expected number of strokes to complete the hole?

14. A promoter thinks that 50,000 fans will attend a game if it rains and 80,000 if it does not rain. What is the expected number of attendees if there is a 40% chance of rain?

Use the following table for questions 15–18. Normal Probabilities for $0 \leq Z \leq z_0$.

| $z_0$ | 0.2 | 0.8 | 1.0 | 1.2 | 1.3 | 1.4 | 1.5 | 2.0 | 2.5 | 2.8 | 3.0 | 3.2 |
|---|---|---|---|---|---|---|---|---|---|---|---|---|
| p | .079 | .288 | .341 | .385 | .403 | .419 | .433 | .477 | .494 | .497 | .499 | .499 |

15. Use the normal distribution table to find

    a. $p(0 < Z < 1.5)$

    b. $p(-1.5 \leq Z \leq 2.0)$

    c. $p(1.0 \leq Z \leq 2.5)$

    d. $p(Z \leq 1.0)$

    e. $p(Z \geq 2.5)$

16. If X is normally distributed with mean 15 and standard deviation 5, find

    a. $p(5 \leq X \leq 15)$

    b. $p(X \geq 5)$

    c. $p(20 < X < 27.5)$

    d. $p(X \leq 10)$

    e. $p(X = 15)$

17. There are 25 true/false questions on a test. Suppose that a student guesses on all of them.

    a. Find the probability that student gets exactly 20 right.

    b. Find the mean and the standard deviation X = the number correct.

    c. Use the normal distribution to approximate the probability that the student gets exactly 20 right.

    d. Use the normal distribution to approximate the probability that the student gets at least 20 right.

18. There are 16 multiple-choice questions on a test. Each question has five possible answers. Suppose that the student guesses on each question.

    a. Find the probability that the student guesses right on exactly 4 questions.

    b. Find the mean and the standard deviation of X = the number correct.

    c. Use the normal distribution to approximate the probability that the student gets 4 right.

    d. Use the normal distribution to approximate the probability that the student gets at most 4 right.

# ✓ Check Yourself

1. a. **(Frequency table)**

| x | 1 | 2 | 3 | 4 | 5 |
|---|---|---|---|---|---|
| f | 3 | 0 | 3 | 1 | 2 |

   b. **(Histogram)**

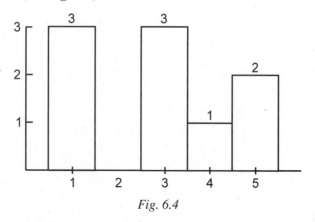

*Fig. 6.4*

c. $\overline{X}$ = the mean = $\dfrac{1(3)+3(3)+4(1)+5(2)}{9} = \dfrac{26}{9} \approx 2.89$. (**Mean**)

d. For the median: the numbers in order are 1, 1, 1, 3, 3, 3, 4, 5, 5 The fifth number is 3, so the median is 3. (**Median**)

e. There are two modes: 1 and 3. (**Mode**)

f. The range is 5 − 1 = 4. (**Range**)

g. The variance:

$$\dfrac{1^2(3) + 2^2(0) + 3^2(3) + 4^2(1) + 5^2(2) - 9(2.89)^2}{9-1} = \dfrac{3 + 27 + 16 + 50 - 75.17}{8} = \dfrac{20.83}{8} = 2.60$$

(**Variance**)

h. The standard deviation = $\sqrt{2.60} \approx 1.61$. (**Standard deviation**)

2. a. (**Histogram**)

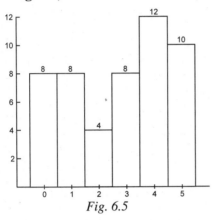

*Fig. 6.5*

b. Mean = $\dfrac{5(10) + 4(12) + 3(8) + 2(4) + 1(8) + 0(8)}{10 + 12 + 8 + 4 + 8 + 8} = \dfrac{130}{50} = 2.6$. (**Mean**)

c. There are 50 numbers, so the median is the average of the 25th and 26th numbers. There are 10 5s, then 12 4s, so the 22nd datum is a 4. Next come 8 3s, so the 23rd to the 30th data are 3s. Thus, the 25th datum is 3, and the 26th datum is 3. The median = $\dfrac{3+3}{2} = 3$. (**Median**)

d. The mode is 4. (**Mode**)

e. The range is 5 − 0 = 5. (**Range**)

f. The variance = $\dfrac{5^2(10) + 4^2(12) + 3^2(8) + 2^2(4) + 1^2(8) + 0^2(8) - 50(2.6)^2}{50-1} = \dfrac{200}{49} \approx 4.08$. (**Variance**)

g. The standard deviation = $\sqrt{4.08} \approx 2.02$. (**Standard deviation**)

3. a. The frequency table is

| x | 10 | 11 | 12 | 13 | 14 | 15 | 16 |
|---|----|----|----|----|----|----|----|
| f | 5 | 12 | 15 | 10 | 4 | 5 | 4 |

(**Frequency tables**)

b. Mean $= \dfrac{10(5) + 11(12) + 12(15) + 13(10) + 14(4) + 15(5) + 16(4)}{5 + 12 + 15 + 10 + 4 + 5 + 4} = \dfrac{687}{55} \approx 12.49.$ **(Mean)**

c. There are 55 data all together, so the median is the 28th number. The 1st through the 5th data are 10s; the 6th through the 17th data are 11s, the 18th through the 32nd data are 13s. Thus, the 28th datum is a 13, and the median = 13. **(Median)**

d. The mode is 12, since there are more 12s then any other number. **(Mode)**

e. The range is $16 - 10 = 6$. **(Range)**

f. Since $10^2\,(5) + 11^2\,(12) + 12^2\,(15) + 13^2\,(10) + 14^2\,(4) + 15^2\,(5) + 16^2\,(4) = 8735$, the variance =

$\dfrac{8735 - (55)(687/55)^2}{54} \approx \dfrac{154}{54} \approx 2.85.$ **(Variance)**

g. Standard deviation $= \sqrt{2.85} \approx 1.7$. **(Standard deviation)**

4.  a. The classes are 10–11, 12–13, 14–15, 16–17, 18–19

| x | 10–11 | 12–13 | 14–15 | 16–17 | 18–19 |
|---|---|---|---|---|---|
| f | 6 | 2 | 3 | 4 | 5 |

**(Frequency tables)**

b. **(Histograms)**

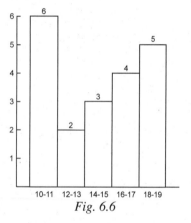

*Fig. 6.6*

c. Mean $\approx \dfrac{10.5(6) + 12.5(2) + 14.5(3) + 16.5(4) + 18.5(5)}{20} = \dfrac{290}{20} = 14.5.$ **(Mean)**

d. The class containing the median is the class that contains the 10th and 11th data. That class is the 14–15 class. **(Median)**

e. The modal class is the class with the most data, which is the 10–11 class. **(Mode)**

f. The range is $19 - 10 = 9$. **(Range)**

g. The variance $\approx \dfrac{10.5^2(6) + 12.5^2(2) + 14.5^2(3) + 16.5^2(4) + 18.5^2(5) - 20(14.5^2)}{19} = \dfrac{200}{19} \approx 10.53.$
**(Variance)**

h. Standard deviation $= \sqrt{10.53} \approx 3.24.$ **(Standard deviation)**

5. a. **(Histograms)**

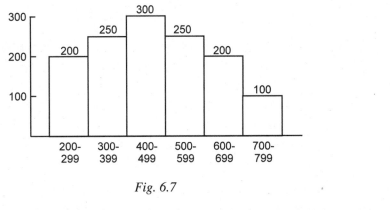

*Fig. 6.7*

b. Mean $\dfrac{250(200) + 350(250) + 450(300) + 550(250) + 650(200) + 750(100)}{1300} = \dfrac{615000}{1300} \approx 473.$ **(Mean)**

c. The class containing the median is the class containing the 650th and 651st data. This class is the 400-499 class. **(Median)**

d. The modal class is the class with the most data, that is, the 400–499 class. **(Mode)**

e. The range $\approx 799 - 200 = 599.$ **(Range)**

f. Since

$250^2\,(200) + 350^2\,(250) + 450^2\,(300) + 550^2\,(250) + 650^2\,(200) + 750^2\,(100) = 320250000,$

Variance $\approx \dfrac{320250000 - 1300(473)^2}{1299} = \dfrac{29402300}{1299} \approx 22635.$ **(Variance)**

g. Standard deviation $\approx \sqrt{22635} \approx 150$ . **(Standard deviation)**

6. a. Frequency table

| x | 20–22 | 23–25 | 26–28 | 29–31 | 32–34 |
|---|---|---|---|---|---|
| f | 10 | 20 | 15 | 5 | 10 |

**(Frequency tables)**

b. Mean $\approx \dfrac{21(10) + 24(20) + 27(15) + 30(5) + 33(10)}{10 + 20 + 15 + 5 + 10} = \dfrac{1575}{60} = 26.25$ **(Mean)**

c. The median is the average of the 30th and 31st data. The 30th datum is in the 23–25 class, and 31st is in the 26–28 class. Thus, the median is $\dfrac{25 + 26}{2} = 25.5.$ **(Median)**

d. The modal class is the 23–25 class. **(Mode)**

e. The range $\approx 34 - 20 = 14.$ **(Range)**

f. Since $21^2\,(10) + 24^2\,(20) + 27^2\,(15) + 30^2\,(5) + 33^2\,(10) = 42255,$

the variance $\approx \dfrac{42255 - 60(26.25)}{59} = \dfrac{911.25}{59} \approx 15.44.$ **(Variance)**

g. Standard deviation $= \sqrt{15.44} \approx 3.93.$ **(Standard deviation)**

7. a. The possible values of X are 0, 1, and 2.

   b. The probability of RR = p(0 G) = = $\frac{3}{10}$ =.3 = $\frac{C(3,2)}{C(5,2)}$.

   The probability of GG = p(2 G) = $\frac{2 \times 1}{5 \times 4}$ =.1. Thus, p(RG or GR) = p(1 G) = 1 −.3 −.1 =.6. The distribution is

   | x | 0 | 1 | 2 |
   |---|---|---|---|
   | p | .3 | .6 | .1 |

   **(Probability distribution)**

   c. Expected value = 0(.3) + 1(.6) + 2(.1) =.8. **(Expected value)**

   d. Variance = $0^2$ (.3) + $1^2$ (.6) + $2^2$ (.1) −$.8^2$ =.36. **(Variance)**

   e. Standard deviation = $\sqrt{.36}$ = .6. **(Standard deviation)**

8. a. The possible values of X are 0, 1, 2, and 3.

   b. X has a binomial distribution, with n = 3 and p =.5. Thus, the probabilities are found using the binomial probability formula. p(0 heads) = C(3, 0)$(.5)^0$ $(.5)^3$ =.125, p(1 head) = C(3, 1)$(.5)^1$ $(.5)^2$ = .375, p(2 heads) = C(3, 2)$(.5)^2$ $(.5)^1$ = .375, and p(3 heads) = C(3, 3)$(.5)^3$ $(.5)^0$ = .125. The distribution is

   | x | 0 | 1 | 2 | 3 |
   |---|---|---|---|---|
   | p | .125 | .375 | .375 | .125 |

   c. Expected value = mean = np = 3(.5) = 1.5

   d. Variance = npq = 3(.5)(.5) =.75

   e. Standard deviation = $\sqrt{.75}$ ≈ .866. **(Binomial distribution)**

9. The probability distribution is

   | x | 1 | 3 | 10 |
   |---|---|---|----|
   | p | .7 | .2 | .1 |

   The expected amount of time per will is

   E = 1(.7) + 3(.2) + 10(.1) = 2.3 hours.

   If she has 80 wills, she should expect to spend 80(2.3) = 184 hours on them. **(Expected value)**

10. a. The distribution is

   | x = winnings | +$3 | +$1 | −$2 |
   |---|---|---|---|
   | p | 1/3 | 1/3 | 1/3 |

   The expected value is E = 3(1/3) + 1(1/3) − 2(1/3) = 2/3. Thus, you expect to win an average of $.67 per game.

   b. The game is not fair (expected value is not zero), and you are favored. **(Expected value)**

11.  a. p(0) = 1 −.3 −.1 −.1 −.05 −.05 = 1 −.6 =.4. **(Probability distribution)**

b. Mean = Expected value

= 0(.4) + 1(.3) + 2(.1) + 4(.04) + 5(.05) =.845. **(Expected value)**

c. Standard deviation = $\sqrt{.3 + 4(.1) + 9(.1) + 25(.05) - .845^2} \approx \sqrt{2.936} \approx 1.71$. **(Standard deviation)**

12.  a.

| x | 0 | 1 | 2 | 3 | 4 | 5 |
|---|---|---|---|---|---|---|
| f | 5 | 30 | 20 | 15 | 7 | 3 |
| p | .0625 | .375 | .25 | .1875 | .0875 | .0375 |

Divide each frequency by 80. **(Probability distribution)**

b.  **(Histograms)**

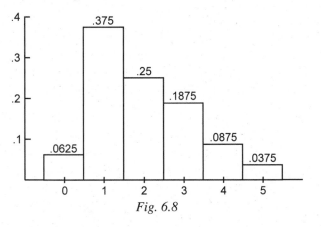

*Fig. 6.8*

13.  a. p(more than 2 strokes) = p(3) + p(4) + p(5) =.25 +.50 +.10 =.85

b. The most common number of strokes is 4 (highest probability). **(Histograms)**

c. E = 1(.05) + 2(.10) + 3(.25) + 4(.50) + 5(.10) = 3.5 strokes. **(Expected value)**

14.

| x | 50000 | 80000 |
|---|---|---|
| p | .4 | .6 |

Expected attendance = 50000(.4) + 80000(.6) = 68,000. **(Expected value)**

15.  a. p(0 < Z < 1.5) = .433

b. p(−1.5 ≤ Z ≤ 2.0) =.477 +.433 = .91

c. p(1.0 ≤ Z ≤ 2.5) =.494 −.341 = .153

d. p(Z ≤ 1.0) =.5 +.341 = .841

e. p(Z ≥ 2.5) =.5 −.494 = .006   **(Standard normal distribution)**

16.  a. $p(5 \leq X \leq 15) = p(\frac{5-15}{5} \leq Z \leq \frac{15-15}{5}) = p(-2 \leq Z \leq 0) = p(0 \leq Z \leq 2) = .477$

b. $p(X \geq 5) = p(Z \geq \frac{5-15}{5}) = p(Z \geq -2) = .477 + .5 = .977$

c. $p(20 < X < 27.5) = p(\frac{20-15}{5} < Z < \frac{27.5-15}{5}) = p(1 < Z < 2.5) = .494 - .341 = .153$

d. $p(X \leq 10) = p(\leq Z) = p(Z \leq -1) = .5 - .341 = .159$

e. $p(X = 15) = 0$  **(Standard normal distribution)**

17.  a. $p(20 \text{ correct}) = C(25, 20)(.5)^{20}(.5)^5 = \frac{25!}{20!5!}(.5)^{25} = \frac{25 \cdot 24 \cdot 23 \cdot 22 \cdot 21}{5 \cdot 4 \cdot 3 \cdot 2 \cdot 1}(.5)^{25} = 53130(.5)^{25} \approx .00158$

b. Mean = $np = 25(.5) = 12.5$, standard deviation = $\sqrt{npq} = \sqrt{25(.5)(.5)} = 2.5$. **(Binomial distribution)**

c. $p(19.5 \leq X \leq 20.5) = p(2.8 \leq Z \leq 3.2) = .499 - .497 = .002$

d. $p(\text{at least } 20 \text{ correct}) = p(X \geq 19.5) = p(Z \geq 2.8) = .5 - .497 = .003$ **(Normal approximation of the binomial)**

18.  a. $p(X = 4) = C(16, 4)(.2)^4(.8)^{12} = \frac{16!}{4!12!}(.2)^4(.8)^{12} = \frac{16 \cdot 15 \cdot 14 \cdot 14}{4 \cdot 3 \cdot 2 \cdot 1}(.2)^4(.8)^{12} = 1820(.2)^4(.8)^{12} \approx .200$

b. Mean = $np = 16(.2) = 3.2$, standard deviation = $\sqrt{npq} = \sqrt{16(.2)(.8)} = \sqrt{2.56} = 1.6$. **(Binomial distribution)**

c. $p(4 \text{ correct}) \approx p(3.5 \leq X \leq 4.5)$

$= p(\frac{3.5-3.2}{1.6} \leq Z \leq \frac{4.5-3.2}{1.6}) = p(.2 \leq Z \leq .8) = .288 - .079 = .209$

d. $p(\text{at most } 4 \text{ right}) = p(X \leq 4.5) = p(Z \leq \frac{4.5-3.2}{1.6}) = p(Z \leq .8) \approx .5 + .288 = .788$ **(Normal approximation to the binomial)**

# Grade Yourself

Circle the numbers of the questions you missed, then fill in the total incorrect for each topic. If you answered more than three questions incorrectly, you need to focus on that topic. (If a topic has less than three questions and you had at least one wrong, we suggest you study that topic also. Read your textbook or a review book, or ask your teacher for help.)

## Subject: Statistics

| Topic | Question Numbers | Number Incorrect |
|---|---|---|
| Frequency table | 1a, 3a, 4a, 6a | |
| Histogram | 1b, 2a, 4b, 5a, 12b, 13a-b | |
| Mean | 1c, 2b, 3b, 4c, 5b, 6b | |
| Median | 1d, 2c, 3c, 4d, 5c, 6c | |
| Mode | 1e, 2d, 3d, 4e, 5d, 6d | |
| Range | 1f, 2e, 3e, 4f, 53e, 6e | |
| Variance | 1g, 2f, 3f, 4g, 5f, 6f, 7d | |
| Standard deviation | 1h. 2g,3g, 4h, 5g, 6g, 7e, 11c | |
| Probability distribution | 7a-b, 9, 10, 11a, 12a, 14 | |
| Expected value | 7c, 9, 10, 11b, 13c, 14 | |
| Binomial distribution | 8a-e, 17a-b, 18a-b | |
| Standard normal distribution | 15, 16 | |
| Normal approximation of the binomial | 17c-d, 18c-d | |

# *Finance*

## Brief Yourself

An arithmetic sequence has the form $a_n = a_1 + (n-1)d$, where $d$ is the common difference between successive terms; the sum of the first $n$ terms of an arithmetic sequence is $s_n = na_1 + \frac{n(n-1)}{2}d$. A geometric sequence has the form $a_n = a_1 r^{n-1}$, where $r$ is the common ratio. The sum of the first $n$ terms of a geometric sequence is $s_n = \frac{a_1(1-r^n)}{1-r}$.

Simple interest on $P$ dollars at interest rate $r$ (expressed as a decimal) over time period $t$ is $I = Prt$. The time periods must match, so that in computing simple interest at 4% per year for 3 months, the 3 months should be changed to 3/12, or 1/4 year. The total amount, or future value, of the principal is $A = P + I = P + Prt = P(1 + rt)$.

If compound interest is used, the amount, or future value, of principal $P$, invested at interest rate $r$ (expressed as a decimal), compounded $m$ times per year for $t$ years is $A = P(1 + i)^n$, where $i = r/m$, the interest rate per compounding period, and $n = mt$, the number of compounding periods. To compare different compound interest rates, we compute the effective interest using the formula $(1 + i)^m - 1$, where $m$ is the number of compounding periods per year, and $i = r/m$. Finally, when interest is compounded continuously, we use the formula $A = Pe^{rt}$ for the future value.

The future value of an ordinary annuity is $A = R \frac{(1+i)^n - 1}{i}$, where R is the periodic payment, $i$ the interest rate per period, and $n$ the number of periods. In an ordinary annuity, the payment is made at the end of each period. The present value of an ordinary annuity is $P = R\frac{(1+i)^n - 1}{i(1+i)^n}$, where $R$ is the payment. This is the formula often used for loans.

# Test Yourself

1.  Decide whether each of the following sequences is arithmetic, geometric, or neither.

    a. 2, 5, 8, 11, 14 ...

    b. 1, -2, 3, -4, 5, -6 ...

    c. $1, \dfrac{1}{2}, \dfrac{1}{3}, \dfrac{1}{4}$ ...

    d. $1, \dfrac{-1}{2}, \dfrac{1}{4}, \dfrac{-1}{8}$ ...

    e. $3, \dfrac{5}{2}, 2, \dfrac{3}{2}, 1, \dfrac{1}{2}$ ...

    f. 2, 6, 18, 54...

2.  Find the formula for each sequence. Then find the 100th term of the sequence, and finally, find the sum of the first 25 terms of the sequence.

    a. 4, 9, 14, 19 ...

    b. 3, –3, –9, –15 ...

3.  Find the formula for each sequence. Then find the 10th term and the sum of the first 10 terms of the sequence.

    a. –2, 20, –200, 2000 ...

    b. $3, \dfrac{3}{4}, \dfrac{3}{16}, \dfrac{3}{64}$ ...

4.  Write out the first six terms of the arithmetic sequence in which $a_1 = -5$, and d = 3.

5.  Write the first six terms of the geometric sequence in which $a_1 = -5$ and r = 3.

6.  A boy is paid 1 cent the 1st day, 2 cents the 2nd day, 4 cents the 3rd day, 8 cents the 4th day, etc.

    a. How much is he paid the 20th day?

    b. How much is he paid altogether for the 20 days?

7.  Find the interest earned if $500 is invested at 5% simple interest for 16 years.

8.  Find the future value of $1200 invested at 6% simple interest for 15 months.

9.  Find the principal that must be invested now at 4.5% simple interest in order to have $3,000 in 9 months.

10. Find the amount that must be invested now at 7% simple interest in order to have $5,000 in 2 years.

11. An investment of $4,000 earned $250 simple interest in 6 months. What was the interest rate?

12. An investment of $200 earns 3% interest compounded monthly. What is the amount in 20 years?

13. What is the value in 10 years of $1,200 invested at 5% compounded quarterly?

14. How much interest does $450 earn at 4% compounded semiannually in 12 years?

15. What is the value in 5 years of $3,000 invested at 6% compounded continuously?

16. How much interest is earned in 3.5 years by $10,000 invested at 4% compounded continuously?

17. How much needs to be invested now at 4.75% compounded continuously in order to have $36,000 in 4 years?

18. You borrow $1,000 for 6 months. How much interest do you pay if interest is 8%

    a. simple interest?

    b. compounded quarterly?

    c. compounded monthly?

    d. continuously compounded?

19. Which interest rate is better: 5.25% compounded monthly or 5.175% compounded daily?

20. Find the effective interest rate of 5.875% compounded monthly.

21. If money can earn 5.5% compounded quarterly, is it better to have $1,000 now or $2,000 in 8 years?

22. What is the future value after 20 years of an ordinary annuity for which the monthly payment is $200 and the interest rate is 4% compounded monthly?

23. You make yearly payments of $500 into an account that earns 3.725% compounded annually. After 30 years, how much interest have you earned?

24. Parents of a newborn child want to have $50,000 towards their child's college education in 18 years. How much should they pay in each month if they can earn 6.5% compounded monthly?

25. John wants $20,000 for a new car in 5 years. What payment should he make if he can receive 4.25% interest compounded monthly?

26. Beth plans to make periodic payments of $25 per week into an account that receives 4.75% interest compounded weekly. If Chavonne wants to have the same amount that Beth has after 12 years, how much should she deposit in a single payment now? Assume she receives 4.75% compounded weekly.

27. Nichelle wants to buy a house. She can afford monthly payments of $750, and she has $10,000 for a down payment. How expensive a house can she afford to buy? Assume that the current 30-year fixed-rate mortgage is 7.25%.

28. Joseph is in the market for a car. He can afford to pay $325 a month. If the dealer offers him $5,000 for his old car, and if the interest rate is 8.25% compounded monthly for 5 years, how expensive a car can Joseph afford?

29. Leroy has $5,000 down payment towards a house that costs $80,000. If the current mortgage rate is 8% for a 30-year fixed-rate loan, what is his monthly payment?

30. In problem 29, assuming that Leroy pays off his 30-year loan, what is the total amount of interest he will pay?

31. If, in problem 29, Leroy can find a bank that will finance his mortgage at 7%, what will his monthly payments be and how much interest will he pay?

32. Joan is arranging to buy a $25,000 van. If the dealership offers her $8,500 trade-in on her old car, what will her monthly payments be if her car loan is 7.75% compounded monthly for 6 years?

33. In problem 32, how much interest will Joan pay on her loan?

34. If Joan can obtain a 6.5% car loan from her credit union for 5 years, what will her payments be? How much interest will she pay? Is this a better deal?

35. Congratulations! You have just won the $6,000,000 lottery. You will receive monthly payments of $25,000 for 20 years. The lottery commission invests some money in an annuity from which these monthly payments will be made. How much money do they need to invest if they can arrange for 10% compounded monthly?

36. Suppose they offer you the option of receiving 20 yearly payments of $300,000. How much would they have to invest in the annuity if the funds receive 10% compounded yearly?

# ✓ Check Yourself

1. a. 2, 5, 8, 11, 14 ... arithmetic: Each term is 3 more than the previous term.

   b. 1, –2, 3, –4, 5, -6 ... neither

   c. $1, \frac{1}{2}, \frac{1}{3}, \frac{1}{4}$ ... neither

   d. $1, \frac{-1}{2}, \frac{1}{4}, \frac{-1}{8}$ ... geometric: Each term is $\frac{-1}{2}$ times the previous term.

   e. $3, \frac{5}{2}, 2, \frac{3}{2}, 1, \frac{1}{2}$ ... arithmetic: Each term is the previous term plus $\frac{-1}{2}$.

   f. 2, 6, 18, 54 ... geometric: Each term is 3 times the previous term. (**Sequences**)

2. a. The formula is $a_n = 4 + (n – 1)5$; $a_{100} = 4 + (99)5 = 499$, $s_{25} = (25)4 + \frac{25(24)}{2}(5) = 100 + 1500 = 1600$.

   b. The formula is $a_n = 4 + (n – 1)(–6)$; $a_{100} = 3 + (99)(–6) = –591$, $s_{25} = 25(3) + \frac{25(24)}{2}(-6) = –1725$.

   (**Sequences**)

3. a. The formula is $a_n = -2(-10)^{n-1}$, $a_{10} = -2(-10)^9 = 2,000,000,000$, and $s_{10} = -2\frac{(1-(-10)^{10})}{1—10} =$
   $-2\frac{1-10000000000}{11} = 2\frac{9999999999}{11} = 2(909090909) = 1818181818$.

   b. The formula is $a_n = 3\left(\frac{1}{4}\right)^{n-1}$, $a_{10} = 3\left(\frac{1}{4}\right)^9 = \frac{3}{262144} \approx 0.0000114$, $s_{10} = 3\frac{1-.25^{10}}{1-.25} = 3\frac{.999999}{.75} \approx$
   3.99999. (**Sequences**)

4. –5, –2, 1, 4, 7, 10. (**Sequences**)

5. –5, –15, –45, –135, –405, –1215. (**Sequences**)

6. a. $a_n = 2^{n-1}$, so that $a_{20} = 2^{19} = 524288$ cents = \$5242.88

   b. $s_{20} = 1\frac{1-2^{20}}{1-2} = \frac{-1048575}{-1} = 1048575$ cents, or \$10485.75. (**Sequences**)

7. $I = Prt = 500(.05)(16) = \$400$. (**Simple interest**)

8. $A = P(1 + rt) = 1200(1 + .06(15/12)) = \$1,290$. (**Simple interest**)

9. $A = 3000 = P(1 + rt) = P(1 + .045(9/12)) = P(1.03375)$, so that $P = \frac{3000}{1.03375} = \$2902.06$. (**Simple interest**)

10. $A = 5000 = P(1 + .07(2)) = P(1.14)$, so that $P = \frac{5000}{1.14} = \$4385.96$. (**Simple interest**)

11. $I = 250 = Prt = 4000(r)(6/12) = 2000r$, so that $r = \frac{250}{2000} = .125$, or 12.5%. (**Simple interest**)

12. $A = P(1 + i)^n = 200(1 + \frac{.03}{12})^{(20)(12)} = 200(1.0025)^{240} = \$364.15$. (**Compound interest**)

13. $A = P(1 + i)^n = 1200(1 + \frac{.05}{4})^{10(4)} = 1200(1.0125)^{40} = \$1972.34$. (**Compound interest**)

14. $A = 450(1 + \frac{.04}{2})^{12(2)} = 450(1.02)^{24} = \$723.80$. $I = A - P = \$723.80 - \$450.00 = \$273.80$. (**Compound interest**)

15. $A = 3000e^{(.06)(5)} = 3000e^{.3} = \$4,049.58$. (**Compound interest**)

16. $A = 10000e^{(.04)(3.5)} = 10000e^{.14} = \$11,502.74$. (**Compound interest**)

17. $A = 36000 = Pe^{(.0475)(4)} = Pe^{.19} \approx P(1.2092496)$, so that $P = \frac{3600}{1.2092496} = \$29,770.53$. (**Compound interest**)

18. You borrow \$1000 for 6 months. How much interest do you pay if interest is 8%

    a. $I = Prt = 1000(.08)(6/12) = \$40$. (**Simple Interest**)

    b. $A = 1000(1 + \frac{.08}{4})^2 = \$1,040.40$, so that $I = A - P = \$1040.40 - \$1000 = \$40.40$.

    c. $A = 1000(1 + \frac{.08}{12})^6 = \$1,040.67$, so that $I = \$1040.67 - \$1000 = \$40.67$.

    d. $A = 1000e^{(.08)(6/12)} = \$1040.81$, so that $I = \$1040.81 - \$1000 = \$40.81$. (**Compound interest**)

19. Effective rate for 5.25% compounded monthly:

    $(1 + .0525/12)^{12} - 1 = .05378$, or 5.378%

    Effective rate for 5.175 compounded daily:

    $(1 + .05175/365)^{365} - 1 = .05310$, or 5.31% 5.25% compounded monthly is the better rate. (**Compound interest**)

20. Effective rate: $(1 + .05875/12)^{12} - 1 = .06038$, or 6.04%. (**Compound interest**)

21. $A = P(1 + i)^n = 1000(1 + .055/12)^{(12)(8)} = \$1,551.15$. Thus, \$1,000 now would be worth \$1,551.15 in 8 years; \$2,000 in 8 years is worth more. (**Compound interest**)

22. $A = R\frac{(1 + i)^n - 1}{i} = 200\frac{(1 + .04/12)^{240} - 1}{(.04/12)} = \$73,354.93$, rounded. (**Annuities**)

23. $A = 500\frac{(1 + .03725)^{30} - 1}{(.03725)} = \$26,788.38$. You have paid in \$500(30) = \$15,000 (30 payments of \$500 each). Thus, $I = \$26,788.38 - \$15,000 = \$11,788.38$. (**Annuities**)

24. $A = 50000 = R\frac{(1 + i)^n - 1}{i} = R\frac{(1 + .065/12)^{216} - 1}{(.065/12)} \approx R(408.33891)$, so that $R = \frac{50000}{408.33891} = \$122.45$. (**Annuities**)

25. $A = 20000 = R\dfrac{(1+i)^n - 1}{i} = R\dfrac{(1 + .045/12)^{60} - 1}{(.045/12)} \approx R(67.14555)$, so that $R = \dfrac{20000}{67.14555} = \$297.86$.

    **(Annuities)**

26. $A = R\dfrac{(1+i)^n - 1}{i} = 25\dfrac{(1 + .0475/52)^{624} - 1}{(.0475/52)} = \$21013.67 =$ Beth's future amount. For Chavonne, $A =$

    $21013.67 = P(1+i)^n = P(1 + .0475/52)^{624} = P(1.7678073)$, so that $P = \dfrac{21013.67}{1.7678073} = \$11,886.86$.

    **(Compound interest and annuities)**

27. $P = R\dfrac{(1+i)^n - 1}{i(1+i)^n} = 750\dfrac{(1 + .0725/12)^{360} - 1}{(.0725/12)(1 + .0725/12)^{60}} = \$109942.26$. Nichelle can afford to pay $\$10,000 +$

    $\$109,942.26$, or about a $\$120,000$ house. **(Loans and amortization)**

28. $P = R\dfrac{(1+i)^n - 1}{i(1+i)^n} = 325\dfrac{(1 + .0825/12)^{60} - 1}{(.0825/12)(1 + .0825/12)^{60}} = \$15,934.30$. Joseph can afford to pay $\$5,000 + \$15,934.30$

    $= \$20,934.30$, or about $\$21,000$ for a car. **(Loans and amortization)**

29. Since he has a $\$5,000$ down payment, the loan amount is

    $A = \$80,000 - \$5,000 = \$75,000 = R\dfrac{(1+i)^n - 1}{i(1+i)^n} = R\dfrac{(1 + .08/12)^{360} - 1}{(.08/12)(1 + .08/12)^{360}} = R(136.28349)$, so that $R =$

    $\dfrac{7500}{136.28349} = \$550.32$. **(Loans and amortization)**

30. Leroy will make 360 payments of $\$550.32$ each, for a total of $\$198,115.20$. Since the amount of the loan is $\$75,000$, the interest is $\$123,115.20$. **(Loans and amortization)**

31. $A = 75000 = R\dfrac{(1+i)^n - 1}{i(1+i)^n} = R\dfrac{(1 + .07/12)^{360} - 1}{(.07/12)(1 + .07/12)^{360}} = R(150.30757)$, so that $R = \dfrac{7500}{150.30757} = \$498.98$.

    Leroy will then pay $(360)(\$498.98) = \$179,632.80$ for the house, and the interest will be $\$104,632.80$. Notice that he saves $\$18,482.40$ in interest. **(Loans and amortization)**

32. Since she has a trade in, the loan amount is $A = 25000 - 8500 = 16500 = R\dfrac{(1+i)^n - 1}{i(1+i)^n} =$

    $R\dfrac{(1 + .0775/12)^{72} - 1}{(.0775/12)(1 + .0775/12)^{72}} = R(57.433554)$, so that $R = \dfrac{16500}{57.433554} = \$287.29$. **(Loans and amortization)**

33. Joan makes 72 payments of $\$287.29$ each, for a total of $\$20,684.88$. Since the principal is $\$16,500$, the interest is $\$20,684.88 - \$16,500 = \$4,184.88$. **(Loans and amortization)**

34. $A = 16500 = R\dfrac{(1+i)^n - 1}{i(1+i)^n} = R\dfrac{(1 + .065/12)^{60} - 1}{(.065/12)(1 + .065/12)^{60}} = R(51.10868)$, so that $R = \dfrac{16500}{51.10868} = \$322.84$. Since

    she will pay $(60)(\$322.84) = \$19,370.40$ for the van, she will pay $\$19,370.40 - \$16,500 = \$2,870.40$ in interest. This certainly looks like a better arrangement, as long as Joan can afford the higher monthly payments. **(Loans and amortization)**

35.  $P = R\dfrac{(1+i)^n - 1}{i(1+i)^n} = 25000\dfrac{(1 + \cdot{}^{10}\!/_{12})^{240} - 1}{(\cdot{}^{10}\!/_{12})\,(1 + \cdot{}^{10}\!/_{12})^{240}} = \$2,590,615.47.$ **(Loans and amortization)**

36.  $P = R\dfrac{(1+i)^n - 1}{i(1+i)^n} = 300000\dfrac{(1 + .1)^{20} - 1}{(.1)\,(1 + .1)^{20}} = \$2,554,069.12.$ **(Loans and amortization)**

# Grade Yourself

Circle the numbers of the questions you missed, then fill in the total incorrect for each topic. If you answered more than three questions incorrectly, you need to focus on that topic. (If a topic has less than three questions and you had at least one wrong, we suggest you study that topic also. Read your textbook or a review book, or ask your teacher for help.)

## Subject: Finance

| Topic | Question Numbers | Number Incorrect |
|---|---|---|
| Sequences | 1, 2, 3, 4, 5, 6 | |
| Simple interest | 7, 8, 9, 10, 11, 18a | |
| Compound interest | 12, 13, 14, 15, 16, 17, 18b-d, 19, 20, 21, 26 | |
| Annuities | 22, 23, 24, 25, 26 | |
| Loans and amortization | 27, 28, 29, 30, 31, 32, 33, 34, 35, 36 | |

# *Markov Processes*

**8**

---

Assume the same experiment is performed over and over, and that the probability of any outcome on the current trial depends at most on the outcome of the previous trial. In this case, a transition matrix governing the process can be written down. Each column of this matrix is an assignment of probabilities, so that the columns add up to 1, and all entries are between 0 and 1 (inclusive). Any square matrix satisfying these two properties is a *stochastic matrix*.

Write down the transition matrix using the format

$$P = \text{to state} \quad \begin{array}{c} \\ 1 \\ 2 \\ 3 \end{array} \overset{\text{from state}}{\overset{1\ 2\ 3}{\begin{bmatrix} & & \\ & & \\ & & \end{bmatrix}}}$$

Starting with a distribution vector, d, we can compute the probabilities of being in the various states after n trials by computing $P^n d$.

A stochastic matrix, P, is regular if some power of P has all positive entries. If P is regular, then the matrices $P^n$ approach a stable matrix; the columns of the stable matrix are the same and give the limiting distribution of the process; to find the stable matrix, solve Px = x, or (I - P)x = 0 for x. All columns of the limiting matrix will equal x, where the sum of the entries of x is 1.

A markov process has an absorbing state if there is a state that always leads back to itself. The column under an absorbing state has all zeros except for one 1 in the row corresponding to that column. For processes with absorbing states, the transition matrix is set up as follows:

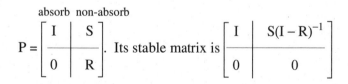

$$P = \begin{bmatrix} I & S \\ \hline 0 & R \end{bmatrix}. \text{ Its stable matrix is } \begin{bmatrix} I & S(I-R)^{-1} \\ \hline 0 & 0 \end{bmatrix}$$

# Test Yourself

1. Which of the following is a stochastic matrix, and how can you tell? In case it is stochastic, is the matrix regular? How can you tell? Does it have any absorbing states? How can you tell?

   a. $\begin{bmatrix} .2 & 1.1 \\ .8 & -.1 \end{bmatrix}$

   b. $\begin{bmatrix} 1 & 0 \\ 0 & 1 \end{bmatrix}$

   c. $\begin{bmatrix} .1 & .3 & .6 \\ .2 & .4 & .2 \\ .6 & .1 & .3 \end{bmatrix}$

   d. $\begin{bmatrix} 1 & .5 & .7 \\ 0 & .3 & .3 \\ 0 & .2 & 0 \end{bmatrix}$

   e. $\begin{bmatrix} .2 & .7 & .25 \\ .8 & .3 & .75 \end{bmatrix}$

   f. $\begin{bmatrix} .2 & .4 & .3 \\ .1 & .6 & .3 \\ .7 & 0 & .4 \end{bmatrix}$

2. A college fundraiser finds that if an alumnus gave to the college the previous year, the probability is .8 that he will give the next year, but if he did not give the previous year, then the probability of his giving the next year is only .25.

   a. Write down the transition matrix. Is the matrix regular? How can you tell? Does it have any absorbing states? How can you tell?

   b. If a person does not give to his college, what is the probability he will give in two years? Three years?

   c. Overall, what percentage of alumni does the fundraiser expect to contribute each year?

3. A pitcher finds that if he struck out the previous batter, the probability is .75 that he will strike the next one out, .10 that the next one will walk, and .15 that the next one will get a run. If the previous batter walked, the probability of striking the next one out is .5, of walking, .25, and of a run, .25. Finally, if the last batter got a

run, the probabilities are .3 for a strikeout, .3 for a walk, and .4 for a run.

   a. Write down the transition matrix. Is the matrix regular? How can you tell? Does it have any absorbing states? How can you tell?

   b. If the initial distribution is d = $\begin{bmatrix} .1 \\ .1 \\ .8 \end{bmatrix}$, then what is the probability that the second better will walk?

   c. Overall, what is this pitcher's percentage of strikeouts, walks, and runs?

4. A gambler bets until either he runs out of money or until he reaches his goal of $3. Each time he bets $1; the probability of winning is .45, and that of losing is .55.

   a. Write down the transition matrix. Is the matrix regular? How can you tell? Does it have any absorbing states? How can you tell?

   b. If the gambler starts with $2, what is the probability that he will have reached his goal in 3 bets?

   c. Find the stable matrix for this process. Explain what it tells you.

5. In a four-year school, 70% of freshmen became sophomores; the rest dropped out. 80% of sophomores became juniors; the rest dropped out. 80% of juniors became seniors; the rest droped out. 90% of the seniors graduated.

   a. Write down the transition matrix. Is the matrix regular? How can you tell? Does it have any absorbing states? How can you tell?

   b. What is the probability of graduating in 4 years, starting as a freshman?

   c. Find the stable matrix and explain what it tells you.

6. There are 4 blocks in a straight line. A person is equally likely to move to any adjoining block.

   a. Write down the transition matrix. Is the matrix regular? How can you tell? Does it have any absorbing states? How can you tell?

   b. If the person starts in the rightmost block, what is the probability that he will reach the leftmost block in 3 moves? 4 moves?

c. Find the stable matrix. What is the long-range probability of being in each of the 4 blocks?

7. There are four blocks in a row. A person is equally likely to move to any adjoining block or stay put.

   a. Write down the transition matrix. Is the matrix regular? How can you tell? Does it have any absorbing states? How can you tell?

   b. If the person starts in the rightmost block, what is the probability that he will reach the leftmost block in 3 moves?

   c. Find the stable matrix. What is the long-range probability of being in each of the 4 blocks?

8. There are four blocks in a row. A person is equally likely to move to one of the adjoining blocks or stay where he is, except that when the leftmost one is reached, he is home and he stays there.

   a. Write down the transition matrix. Is the matrix regular? How can you tell? Does it have any absorbing states? How can you tell?

b. If the person starts in the rightmost block, what is the probability that he will reach the leftmost block in 3 moves?

c. Find the stable matrix. What is the long-range probability of being in each of the 4 blocks?

9. There are four blocks in a row. A person is equally likely to move to an adjoining block or stay put, except that when either the leftmost one or the rightmost one is reached, the person stays there.

   a. Write down the transition matrix. Is the matrix regular? How can you tell? Does it have any absorbing states? How can you tell?

   b. If the person starts in the right most block, what is the probability that he will reach the left most block in 3 moves?

   c. Find the stable matrix. What is the long-range probability of being in each of the 4 blocks?

---

# ✔ Check Yourself

1.  a. $\begin{bmatrix} .2 & 1.1 \\ .8 & -.1 \end{bmatrix}$ is not stochastic because not all entries are positive.

   b. $\begin{bmatrix} 1 & 0 \\ 0 & 1 \end{bmatrix}$ is stochastic because (1) it is square, (2) all entries are between 0 and 1, inclusive, and (3) each column sums to 1. It is not regular, since all powers of the matrix look like the first power, and that has non-positive entries. Both states are absorbing; we can tell because there are 1s down the main diagonal.

   c. $\begin{bmatrix} .1 & .3 & .6 \\ .2 & .4 & .2 \\ .6 & .1 & .3 \end{bmatrix}$ is not stochastic because the first column does not add up to 1. (All columns should sum to 1.)

   d. $\begin{bmatrix} 1 & .5 & .7 \\ 0 & .3 & .3 \\ 0 & .2 & 0 \end{bmatrix}$ is stochastic because (1) it is square, (2) all entries are between 0 and 1, inclusive, and (3) each column sums to 1. It is not regular, since all powers of the matrix have the same first column. State 1 is an absorbing state; we can tell because there is a 1 in the (1, 1) position.

   e. $\begin{bmatrix} .2 & .7 & .25 \\ .8 & .3 & .75 \end{bmatrix}$ is not stochastic because it is not square. (**Stochastic matrices**)

f. $\begin{bmatrix} .2 & .4 & .3 \\ .1 & .6 & .3 \\ .7 & 0 & .4 \end{bmatrix}$ = A is stochastic because (1) it is square, (2) all entries are between 0 and 1, inclusive, and (3)

each column sums to 1. $A^2 = \begin{bmatrix} .29 & .32 & .30 \\ .29 & .40 & .33 \\ .42 & .28 & .37 \end{bmatrix}$ (all entries positive), so A is regular. There are no absorbing

states because there are no 1s on the main diagonal. (**Transition matrices**)

2.

       Gives  No Gift

a. $P = \begin{matrix} \text{Gives} \\ \text{No Gift} \end{matrix} \begin{bmatrix} .8 & .25 \\ .2 & .75 \end{bmatrix}$. This matrix is regular, since all entries are positive. There are no absorbing states,

since there are no 1s on the main diagonal. (**Transition matrix**)

b. $p^2 \begin{bmatrix} 0 \\ 1 \end{bmatrix} = \begin{bmatrix} .69 & .3875 \\ .31 & .6125 \end{bmatrix} \begin{bmatrix} 0 \\ 1 \end{bmatrix}, \begin{bmatrix} .3875 \\ .6125 \end{bmatrix}$ so the probability of giving after 2 years is .3875, if the person did

not give initially. $p^3 \begin{bmatrix} 0 \\ 1 \end{bmatrix} = \begin{bmatrix} .6925 & .463125 \\ .3705 & .536875 \end{bmatrix} \begin{bmatrix} 0 \\ 1 \end{bmatrix} = \begin{bmatrix} .463125 \\ .536875 \end{bmatrix}$, so the probability of giving after 3 years is

.463125. (**Transition matrix**)

c. The stable matrix can be found by solving (I - P)X = 0, sum of the entries of X = 1.

$(I - P) = \begin{bmatrix} .2 & -.25 \\ -.2 & .25 \end{bmatrix} \rightarrow \begin{bmatrix} .2 & -.25 \\ 0 & 0 \end{bmatrix} \rightarrow \begin{bmatrix} 1 & -1.25 \\ 0 & 0 \end{bmatrix}$, so that x - 1.25y = 0, or x = 1.25y, or $X = \begin{bmatrix} x \\ y \end{bmatrix} = \begin{bmatrix} 1.25t \\ t \end{bmatrix}$,

and 1.25t + t = 2.25t = 1. Thus, $t = \dfrac{1}{2.25} = \dfrac{4}{9} \approx .44$, and $X = \begin{bmatrix} 5/9 \\ 4/9 \end{bmatrix} \approx \begin{bmatrix} .56 \\ .44 \end{bmatrix}$. Thus, the stable matrix is

$\begin{bmatrix} .56 & .56 \\ .44 & .44 \end{bmatrix}$.

Overall, about 56% will give each year. (**Regular stable matrix**)

3. A pitcher finds that if he struck out the previous batter, the probability is .75 that he will strike the next one out, .10 that the next one will walk, and .15 that the next one will get a run. If the previous batter walked, the probability of striking the next one out is .5, of walking, .25, and of a run, .25. Finally, if the last batter got a run, the probabilities are .3 for a strikeout, .3 for a walk, and .4 for a run.

        S  W  R

a. $P = \begin{matrix} S \\ W \\ R \end{matrix} \begin{bmatrix} .75 & .50 & .30 \\ .10 & .25 & .30 \\ .15 & .25 & .40 \end{bmatrix}$ is regular since all entries are positive. There are no absorbing states (no 1s down

the main diagonal). (**Transition matrices**)

b. $p^2 \begin{bmatrix} .1 \\ .1 \\ .8 \end{bmatrix} = \begin{bmatrix} .6575 & .5750 & .495 \\ .1450 & .1875 & .225 \\ .1975 & .2375 & .280 \end{bmatrix} \begin{bmatrix} .1 \\ .1 \\ .8 \end{bmatrix} = \begin{bmatrix} .519 \\ .213 \\ .268 \end{bmatrix}$. The probability is .213 for a walk. (**Transition matrices**)

c. Find the stable matrix by solving (I - P)X = 0, entries of X sum to 1.

$(I - P) = \begin{bmatrix} .25 & -.50 & -.30 \\ -.10 & .75 & -.30 \\ -.15 & -.25 & .60 \end{bmatrix} \rightarrow \begin{bmatrix} 1 & -2 & -1.2 \\ -.10 & .75 & -.30 \\ -.15 & -.25 & .60 \end{bmatrix} \rightarrow \begin{bmatrix} 1 & -2 & -2 \\ 0 & .55 & -.50 \\ -.15 & -.25 & .60 \end{bmatrix} \rightarrow \begin{bmatrix} 1 & -2 & -2 \\ 0 & .55 & -4.2 \\ -.15 & -.25 & .60 \end{bmatrix} \rightarrow$

$\begin{bmatrix} 1 & -2 & -2 \\ 0 & .55 & -4.2 \\ 0 & -.55 & 4.2 \end{bmatrix} \rightarrow \begin{bmatrix} 1 & -2 & -2 \\ 0 & .55 & -4.2 \\ 0 & 0 & 0 \end{bmatrix} \rightarrow \begin{bmatrix} 1 & -2 & -2 \\ 0 & 1 & -420/55 \\ 0 & 0 & 0 \end{bmatrix} \rightarrow \begin{bmatrix} 1 & 0 & -30/11 \\ 0 & 1 & -420/55 \\ 0 & 0 & 0 \end{bmatrix}$, so that $X = \begin{bmatrix} 30/11t \\ 420/55t \\ t \end{bmatrix}$, and $\dfrac{30}{11}t + \dfrac{420}{55}t + t$

$$= \frac{150+420+55}{55}t = \frac{125}{11}t = 1, \text{ thus, } t = \frac{11}{125} \approx .088, \text{ and } X \approx \begin{bmatrix} .240 \\ .672 \\ .088 \end{bmatrix}, \text{ and the stable matrix is } \begin{bmatrix} .24 & .24 & .24 \\ .67 & .67 & .67 \\ .09 & .09 & .09 \end{bmatrix}.$$

Thus, overall , about 24% strike out, 67% walk, and 9% get runs. (**Regular stable matrices**)

4. a.
$$P = \begin{array}{c} \\ 0 \\ 1 \\ 2 \\ 3 \end{array} \begin{array}{cccc} 0 & 1 & 2 & 3 \\ \begin{bmatrix} 1 & .55 & 0 & 0 \\ 0 & 0 & .55 & 0 \\ 0 & .45 & 0 & 0 \\ 0 & 0 & .45 & 1 \end{bmatrix} \end{array}.$$

This matrix has two absorbing states, $0 and $3. There are 1s in the main diagonal in the (1, 1) and (4, 4) positions. Thus, the matrix is not regular. Normally, the matrix is rearranged with the absorbing states first.

$$P = \begin{array}{c} \\ 0 \\ 1 \\ 2 \\ 3 \end{array} \begin{array}{cccc} 0 & 3 & 1 & 2 \\ \begin{bmatrix} 1 & 0 & .55 & 0 \\ 0 & 1 & 0 & .45 \\ 0 & 0 & 0 & .55 \\ 0 & 0 & .45 & 0 \end{bmatrix} \end{array}. \text{ Notice the new order of states. } (\textbf{Absorbing processes})$$

b. $P^3 \begin{bmatrix} 0 \\ 0 \\ 0 \\ 1 \end{bmatrix} = \begin{bmatrix} 1 & 0 & .686125 & .302500 \\ 0 & 1 & .202500 & .561375 \\ 0 & 0 & 0 & .136125 \\ 0 & 0 & .111375 & 0 \end{bmatrix} \begin{bmatrix} 0 \\ 0 \\ 0 \\ 1 \end{bmatrix} = \begin{bmatrix} .302500 \\ .511375 \\ .136125 \\ 0 \end{bmatrix},$ so the probability of having $3 after 3 bets is

.511375, or about .51. (**transition matrices**)

c. To find the stable matrix when $P = \left[\begin{array}{c|c} I & S \\ \hline 0 & R \end{array}\right]$, find $\left[\begin{array}{c|c} I & S(I-R)^{-1} \\ \hline 0 & R \end{array}\right]$. $S(I - R)^{-1} = \begin{bmatrix} .55 & 0 \\ 0 & .45 \end{bmatrix}$

$\begin{bmatrix} 1 & -.55 \\ -.45 & 1 \end{bmatrix}^{-1}$.

To find $\begin{bmatrix} 1 & -.55 \\ -.45 & 1 \end{bmatrix}^{-1}$, reduce $\left[\begin{array}{cc|cc} 1 & -.55 & 1 & 0 \\ -.45 & 1 & 0 & 1 \end{array}\right] \rightarrow \left[\begin{array}{cc|cc} 1 & -.55 & 1 & 0 \\ 0 & {}^{301}\!/_{400} & {}^{9}\!/_{20} & 1 \end{array}\right] \rightarrow \left[\begin{array}{cc|cc} 1 & -.55 & 1 & 0 \\ 0 & 1 & {}^{180}\!/_{301} & {}^{400}\!/_{301} \end{array}\right] \rightarrow$

$\left[\begin{array}{cc|cc} 1 & 0 & {}^{400}\!/_{301} & {}^{220}\!/_{301} \\ 0 & 1 & {}^{180}\!/_{301} & {}^{400}\!/_{301} \end{array}\right]$, so that $(I - R)^{-1} = \begin{bmatrix} {}^{400}\!/_{301} & {}^{220}\!/_{301} \\ {}^{180}\!/_{301} & {}^{400}\!/_{301} \end{bmatrix} \approx \begin{bmatrix} 1.3289 & .7309 \\ .5980 & 1.3289 \end{bmatrix}$. Thus, $S(I - R)^{-1} =$

$\begin{bmatrix} .55 & 0 \\ 0 & .45 \end{bmatrix}\begin{bmatrix} 1.3289 & .7309 \\ .5980 & 1.3289 \end{bmatrix} \approx \begin{bmatrix} .73 & .40 \\ .27 & .60 \end{bmatrix}$. Thus, the stable matrix is $\begin{bmatrix} 1 & 0 & .73 & .4 \\ 0 & 1 & .27 & .6 \\ 0 & 0 & 0 & 0 \\ 0 & 0 & 0 & 0 \end{bmatrix}$.

Starting with $1, the gambler has a .73 chance of going to state $0 (broke) and a .27 chance of ending up with $3. Starting with $2, the chance of ending up in state $0 is .4, and the chance of ending up in state $3 is .6. (**Absorbing processes**)

5.

$$
\text{a. } P = \begin{array}{c} \\ D \\ G \\ F \\ S \\ J \\ Sn \end{array}
\begin{array}{c} D\ \ G\ \ F\ \ \ S\ \ \ J\ \ Sn \\
\begin{bmatrix}
1 & 0 & .3 & .2 & .2 & .1 \\
0 & 1 & 0 & 0 & 0 & .9 \\
0 & 0 & 0 & 0 & 0 & 0 \\
0 & 0 & .7 & 0 & 0 & 0 \\
0 & 0 & 0 & .8 & 0 & 0 \\
0 & 0 & 0 & 0 & .8 & 0
\end{bmatrix}
\end{array}
$$, and is not regular since there are two absorbing states, D (dropping out) and

G (graduating). The 1s in the main diagonal indicate that these are absorbing. (**Absorbing processes**)

b. $P^4 \begin{bmatrix} 0 \\ 0 \\ 1 \\ 0 \\ 0 \\ 0 \end{bmatrix} = \begin{bmatrix} 1 & 0 & .5968 & .424 & .28 & .1 \\ 0 & 1 & .4032 & .576 & .72 & .9 \\ 0 & 0 & 0 & 0 & 0 & 0 \\ 0 & 0 & 0 & 0 & 0 & 0 \\ 0 & 0 & 0 & 0 & 0 & 0 \\ 0 & 0 & 0 & 0 & 0 & 0 \end{bmatrix} \begin{bmatrix} 0 \\ 0 \\ 1 \\ 0 \\ 0 \\ 0 \end{bmatrix} = \begin{bmatrix} .5968 \\ .4032 \\ 0 \\ 0 \\ 0 \\ 0 \end{bmatrix}$,

so that after 4 years, the probability of graduating is .4032, or about 40%. (**Transition matrices**)

c. In this case, we have the stable matrix, $\begin{bmatrix} 1 & 0 & .5968 & .424 & .28 & .1 \\ 0 & 1 & .4032 & .576 & .72 & .9 \\ 0 & 0 & 0 & 0 & 0 & 0 \\ 0 & 0 & 0 & 0 & 0 & 0 \\ 0 & 0 & 0 & 0 & 0 & 0 \\ 0 & 0 & 0 & 0 & 0 & 0 \end{bmatrix}$.

The probabilities of graduating starting as a freshman are .40, as a sophomore, .58, as a junior, .72, and as a senior, .9. The probabilities of dropping out are, respectively, .6, .42, .28, and .1. (**Absorbing processes**)

6.

$$
\text{a. } P = \begin{array}{c} \\ 1 \\ 2 \\ 3 \\ 4 \end{array}
\begin{array}{c} 1\ \ 2\ \ 3\ \ 4 \\
\begin{bmatrix}
0 & .5 & 0 & 0 \\
1 & 0 & .5 & 0 \\
0 & .5 & 0 & 1 \\
0 & 0 & .5 & 0
\end{bmatrix}
\end{array}
$$. P has no absorbing states, since there are no 1s on the main diagonal.

$P^2 = \begin{bmatrix} .5 & 0 & .25 & 0 \\ 0 & .75 & 0 & .5 \\ .5 & 0 & .75 & 0 \\ 0 & .25 & 0 & .5 \end{bmatrix}$, $P^3 = \begin{bmatrix} 0 & .3750 & 0 & .25 \\ .75 & 0 & .625 & 0 \\ 0 & .625 & 0 & .75 \\ .25 & 0 & .375 & 0 \end{bmatrix}$, and $P^4 = \begin{bmatrix} .375 & 0 & .31 & 0 \\ 0 & .69 & 0 & .625 \\ .625 & 0 & .69 & 0 \\ 0 & .31 & 0 & .375 \end{bmatrix}$.

Successive powers of P have half their entries 0, and the non-zero entries occur in the places where the previous and the next power have 0's. Thus, P cannot be regular. (**Absorbing processes**)

b. Since $P^3 \begin{bmatrix} 0 \\ 0 \\ 0 \\ 1 \end{bmatrix} = \begin{bmatrix} 0 & .3750 & 0 & .25 \\ .75 & 0 & .625 & 0 \\ 0 & .625 & 0 & .75 \\ .25 & 0 & .375 & 0 \end{bmatrix} \begin{bmatrix} 0 \\ 0 \\ 0 \\ 1 \end{bmatrix} = \begin{bmatrix} .25 \\ 0 \\ .75 \\ 0 \end{bmatrix}$, the probability of reaching the leftmost block in

3 moves, starting in the rightmost block is .25. $P^4 \begin{bmatrix} 0 \\ 0 \\ 0 \\ 1 \end{bmatrix} = \begin{bmatrix} .375 & 0 & .31 & 0 \\ 0 & .69 & 0 & .625 \\ .625 & 0 & .69 & 0 \\ 0 & .31 & 0 & .375 \end{bmatrix} \begin{bmatrix} 0 \\ 0 \\ 0 \\ 1 \end{bmatrix} = \begin{bmatrix} 0 \\ .625 \\ 0 \\ .375 \end{bmatrix}$,

so that it is not possible to reach the leftmost block from the rightmost block in 4 moves. (**Transition matrices**)

c. This process does not have a stable matrix. You can see this by the patterns of zeros in the successive powers of P. (**Transition matrices**)

7.

$$\begin{array}{cccc} & 1 & 2 & 3 & 4 \end{array}$$

a. $P = \begin{array}{c} 1 \\ 2 \\ 3 \\ 4 \end{array}\begin{bmatrix} \frac{1}{2} & \frac{1}{3} & 0 & 0 \\ \frac{1}{2} & \frac{1}{3} & \frac{1}{3} & 0 \\ 0 & \frac{1}{3} & \frac{1}{3} & \frac{1}{2} \\ 0 & 0 & \frac{1}{3} & \frac{1}{2} \end{bmatrix}$, and P has no absorbing states since there are no 1s on the main diagonal.

$$P^2 = \begin{bmatrix} \frac{5}{12} & \frac{5}{18} & \frac{1}{9} & 0 \\ \frac{5}{12} & \frac{7}{18} & \frac{2}{9} & \frac{1}{6} \\ \frac{1}{6} & \frac{2}{9} & \frac{7}{18} & \frac{5}{12} \\ 0 & \frac{1}{9} & \frac{5}{18} & \frac{5}{12} \end{bmatrix} \approx \begin{bmatrix} .42 & .28 & .11 & 0 \\ .42 & .39 & .22 & .17 \\ .17 & .22 & .39 & .42 \\ 0 & .11 & .28 & .42 \end{bmatrix}, \text{ and } P^3 = \begin{bmatrix} \frac{25}{72} & \frac{29}{108} & \frac{7}{54} & \frac{1}{18} \\ \frac{29}{72} & \frac{37}{108} & \frac{7}{27} & \frac{7}{36} \\ \frac{7}{36} & \frac{7}{27} & \frac{37}{108} & \frac{29}{72} \\ \frac{1}{18} & \frac{7}{54} & \frac{29}{108} & \frac{25}{72} \end{bmatrix}\begin{bmatrix} .35 & .27 & .13 & .06 \\ .40 & .34 & .26 & .19 \\ .19 & .26 & .34 & .40 \\ .06 & .13 & .27 & .35 \end{bmatrix}.$$

Therefore, P is regular. (A power of P, $P^3$, has all positive entries.) (**Regular matrices**)

b. Since $P^3\begin{bmatrix} 0 \\ 0 \\ 0 \\ 1 \end{bmatrix} = \begin{bmatrix} .35 & .27 & .13 & .06 \\ .40 & .34 & .26 & .19 \\ .19 & .26 & .34 & .40 \\ .06 & .13 & .27 & .35 \end{bmatrix}\begin{bmatrix} 0 \\ 0 \\ 0 \\ 1 \end{bmatrix} = \begin{bmatrix} .06 \\ .19 \\ .40 \\ .35 \end{bmatrix}$, the probability of reaching the leftmost block in 3 moves

starting at the rightmost block is .06. (**Transition matrices**)

c. To find the stable matrix, solve (I - P)X = 0, entries of X sum to 1.

$$(I - P) = \begin{bmatrix} \frac{1}{2} & -\frac{1}{3} & 0 & 0 \\ -\frac{1}{2} & \frac{2}{3} & -\frac{1}{3} & 0 \\ 0 & -\frac{1}{3} & \frac{2}{3} & -\frac{1}{2} \\ 0 & 0 & -\frac{1}{3} & \frac{1}{2} \end{bmatrix} \rightarrow \begin{bmatrix} \frac{1}{2} & -\frac{1}{3} & 0 & 0 \\ 0 & \frac{1}{3} & -\frac{1}{3} & 0 \\ 0 & -\frac{1}{3} & \frac{2}{3} & -\frac{1}{2} \\ 0 & 0 & -\frac{1}{3} & \frac{1}{2} \end{bmatrix} \rightarrow \begin{bmatrix} \frac{1}{2} & -\frac{1}{3} & 0 & 0 \\ 0 & \frac{1}{3} & -\frac{1}{3} & 0 \\ 0 & 0 & \frac{1}{3} & -\frac{1}{2} \\ 0 & 0 & -\frac{1}{3} & \frac{1}{2} \end{bmatrix} \rightarrow \begin{bmatrix} \frac{1}{2} & -\frac{1}{3} & 0 & 0 \\ 0 & \frac{1}{3} & -\frac{1}{3} & 0 \\ 0 & 0 & \frac{1}{3} & -\frac{1}{2} \\ 0 & 0 & 0 & 0 \end{bmatrix} \rightarrow$$

$$\begin{bmatrix} \frac{1}{2} & -\frac{1}{3} & 0 & 0 \\ 0 & \frac{1}{3} & 0 & -\frac{1}{2} \\ 0 & 0 & \frac{1}{3} & -\frac{1}{2} \\ 0 & 0 & 0 & 0 \end{bmatrix} \rightarrow \begin{bmatrix} \frac{1}{2} & 0 & 0 & -\frac{1}{2} \\ 0 & \frac{1}{3} & 0 & -\frac{1}{2} \\ 0 & 0 & \frac{1}{3} & -\frac{1}{2} \\ 0 & 0 & 0 & 0 \end{bmatrix} \rightarrow \begin{bmatrix} 1 & 0 & 0 & -1 \\ 0 & 1 & 0 & -\frac{3}{2} \\ 0 & 0 & 1 & -\frac{3}{2} \\ 0 & 0 & 0 & 0 \end{bmatrix}, \text{ so that } X = \begin{bmatrix} t \\ 1.5t \\ 1.5t \\ t \end{bmatrix},$$

where t + 1.5t + 1.5t + t = 5t = 1, so that t = .2 and $X = \begin{bmatrix} .2 \\ .3 \\ .3 \\ .2 \end{bmatrix}$. The stable matrix is $\begin{bmatrix} .2 & .2 & .2 & .2 \\ .3 & .3 & .3 & .3 \\ .3 & .3 & .3 & .3 \\ .2 & .2 & .2 & .2 \end{bmatrix}$.

The long range probability of being in each of the end blocks is .2, and of being in one of the middle blocks, .3. (**Regular stable matrices**)

8. a. $P = \begin{bmatrix} 1 & \frac{1}{3} & 0 & 0 \\ 0 & \frac{1}{3} & \frac{1}{3} & 0 \\ 0 & \frac{1}{3} & \frac{1}{3} & \frac{1}{2} \\ 0 & 0 & \frac{1}{3} & \frac{1}{2} \end{bmatrix}$. P is not regular; the first state, corresponding to the leftmost block, is absorbing.

(**Absorbing matrices**)

b. $P^3 = \begin{bmatrix} 1 & .52 & .19 & .06 \\ 0 & .15 & .20 & .19 \\ 0 & .20 & .34 & .40 \\ 0 & .13 & .27 & .35 \end{bmatrix}$, so the probability of ending up in the leftmost block in 3 moves starting from the

rightmost block is .06. (**Transition matrices**)

c. To find the stable matrix, compute $S(I - R)^{-1} = [\frac{1}{3} \ 0 \ 0] \left\{ \begin{bmatrix} 1 & 0 & 0 \\ 0 & 1 & 0 \\ 0 & 0 & 1 \end{bmatrix} - \begin{bmatrix} \frac{1}{3} & \frac{1}{3} & 0 \\ \frac{1}{3} & \frac{1}{3} & \frac{1}{2} \\ 0 & \frac{1}{3} & \frac{1}{2} \end{bmatrix} \right\}^{-1}$ .

$I - R = \begin{bmatrix} \frac{2}{3} & -\frac{1}{3} & 0 \\ -\frac{1}{3} & \frac{2}{3} & -\frac{1}{2} \\ 0 & -\frac{1}{3} & \frac{1}{2} \end{bmatrix}$ .. $\begin{bmatrix} \frac{2}{3} & -\frac{1}{3} & 0 & | & 1 & 0 & 0 \\ -\frac{1}{3} & \frac{2}{3} & -\frac{1}{2} & | & 0 & 1 & 0 \\ 0 & -\frac{1}{3} & \frac{1}{2} & | & 0 & 0 & 1 \end{bmatrix} \rightarrow \begin{bmatrix} \frac{2}{3} & -\frac{1}{3} & 0 & | & 1 & 0 & 0 \\ 0 & \frac{1}{2} & -\frac{1}{2} & | & .5 & 1 & 0 \\ 0 & -\frac{1}{3} & \frac{1}{2} & | & 0 & 0 & 1 \end{bmatrix} \rightarrow \begin{bmatrix} \frac{2}{3} & -\frac{1}{3} & 0 & | & 1 & 0 & 0 \\ 0 & 1 & -1 & | & 1 & 2 & 0 \\ 0 & 0 & \frac{1}{6} & | & \frac{1}{3} & \frac{2}{3} & 1 \end{bmatrix} \rightarrow$

$\begin{bmatrix} \frac{2}{3} & -\frac{1}{3} & 0 & | & 1 & 0 & 0 \\ 0 & 1 & 0 & | & 3 & 6 & 6 \\ 0 & 0 & 1 & | & 2 & 4 & 6 \end{bmatrix} \rightarrow \begin{bmatrix} \frac{2}{3} & 0 & 0 & | & 2 & 2 & 2 \\ 0 & 1 & 0 & | & 3 & 6 & 6 \\ 0 & 0 & 1 & | & 2 & 4 & 6 \end{bmatrix} \rightarrow \begin{bmatrix} 1 & 0 & 0 & | & 3 & 3 & 3 \\ 0 & 1 & 0 & | & 3 & 6 & 6 \\ 0 & 0 & 1 & | & 2 & 4 & 6 \end{bmatrix}$ , so $(I-R)^{-1} = \begin{bmatrix} 3 & 3 & 3 \\ 3 & 6 & 6 \\ 2 & 4 & 6 \end{bmatrix}$ .

Thus, $S(I - R)^{-1} = [\frac{1}{3} \ 0 \ 0] \begin{bmatrix} 3 & 3 & 3 \\ 3 & 6 & 6 \\ 2 & 4 & 6 \end{bmatrix} = [ 1 \ 1 \ 1 ]$ , and the stable matrix is $\begin{bmatrix} 1 & 1 & 1 & 1 \\ 0 & 0 & 0 & 0 \\ 0 & 0 & 0 & 0 \\ 0 & 0 & 0 & 0 \end{bmatrix}$ . This means that in the

long run, the person always ends up in the leftmost block. However, we could have saved ourselves the computation. With only one absorbing state, and with the possibility of reaching that state from any of the others (in some number of moves), this is the only possibility. (**Absorbing processes**)

9. $P = \begin{bmatrix} 1 & \frac{1}{3} & 0 & 0 \\ 0 & \frac{1}{3} & \frac{1}{3} & 0 \\ 0 & \frac{1}{3} & \frac{1}{3} & 0 \\ 0 & 0 & \frac{1}{3} & 1 \end{bmatrix}$ . There are two absorbing states, landing in the leftmost and the rightmost blocks.

$$\begin{array}{c} \quad\quad 1 \ 4 \ 2 \ 3 \\ \begin{array}{c} 1 \\ 4 \\ 2 \\ 3 \end{array} \begin{bmatrix} 1 & 0 & \frac{1}{3} & 0 \\ 0 & 1 & 0 & \frac{1}{3} \\ 0 & 0 & \frac{1}{3} & \frac{1}{3} \\ 0 & 0 & \frac{1}{3} & \frac{1}{3} \end{bmatrix} \end{array}$$

Reorder the states, 1 4 2 3 and write $P = $

b. It is not possible to reach the leftmost block starting at the rightmost block, since state 4 (right block) is absorbing. Starting in this block, a person can only stay in this block.

c. To find the stable matrix, compute $S(I - R)^{-1}$. $I - R = \begin{bmatrix} \frac{2}{3} & -\frac{1}{3} \\ -\frac{1}{3} & \frac{2}{3} \end{bmatrix}$ . Reduce $\begin{bmatrix} \frac{2}{3} & -\frac{1}{3} & | & 1 & 0 \\ -\frac{1}{3} & \frac{2}{3} & | & 0 & 1 \end{bmatrix} \rightarrow$

$\begin{bmatrix} 2 & -1 & | & 3 & 0 \\ -1 & 2 & | & 0 & 3 \end{bmatrix} \rightarrow \begin{bmatrix} -1 & 2 & | & 0 & 3 \\ 2 & -1 & | & 3 & 0 \end{bmatrix} \rightarrow \begin{bmatrix} -1 & 2 & | & 0 & 3 \\ 0 & 3 & | & 3 & 6 \end{bmatrix} \rightarrow \begin{bmatrix} 1 & -2 & | & 0 & -3 \\ 0 & 1 & | & 1 & 2 \end{bmatrix} \rightarrow \begin{bmatrix} 1 & 0 & | & 2 & 1 \\ 0 & 1 & | & 1 & 2 \end{bmatrix}$ .

Thus, $S(I - R) \begin{bmatrix} \frac{1}{3} & 0 \\ 0 & \frac{1}{3} \end{bmatrix} \begin{bmatrix} 2 & 1 \\ 1 & 2 \end{bmatrix} = \begin{bmatrix} \frac{2}{3} & \frac{1}{3} \\ \frac{1}{3} & \frac{2}{3} \end{bmatrix}$ , and the stable matrix is $\begin{bmatrix} 1 & 0 & \frac{2}{3} & \frac{1}{3} \\ 0 & 1 & \frac{1}{3} & \frac{2}{3} \\ 0 & 0 & 0 & 0 \\ 0 & 0 & 0 & 0 \end{bmatrix}$ . Starting from state 2,

there is a 2/3 probability of ending up in the leftmost block and a 1/3 probability of ending in the rightmost. These probabilities are reversed if the person starts in block 3. In the long range, the probability is zero of ending up in either of the middle blocks. Thus, the long-range probability of ending up in the left most block is $\frac{1}{4}(1) + \frac{1}{4}\left(\frac{2}{3}\right) + \frac{1}{4}\left(\frac{1}{3}\right) + \frac{1}{4}(0) = \frac{1}{2}$. The probability of ending up in the right most block is

$\frac{1}{4}(0) + \frac{1}{4}\left(\frac{1}{3}\right) + \frac{1}{4}\left(\frac{2}{3}\right) + \frac{1}{4}(1) = \frac{1}{2}$. The probability of starting in any particular block is $\frac{1}{4}$. (**Absorbing processes**)

# Grade Yourself

Circle the numbers of the questions you missed, then fill in the total incorrect for each topic. If you answered more than three questions incorrectly, you need to focus on that topic. (If a topic has less than three questions and you had at least one wrong, we suggest you study that topic also. Read your textbook or a review book, or ask your teacher for help.)

## Subject: Markov Processes

| Topic | Question Numbers | Number Incorrect |
| --- | --- | --- |
| Stochastic/transition matrices | 1a-e, | |
| Transition matrices | 1f, 2a-b, 3a-b, 4b, 5b, 6b-c, 7b, 8b | |
| Regular matrix | 2c, 3c, 7a, 7c | |
| Absorbing processes | 4a, 4c, 5a, 5c, 6a, 8a, 8c, 9a-c | |

# *Game Theory*

## 9

## Brief Yourself

A two-person zero-sum game is a game with two players, R and C, such that each gain for R is a loss of the same amount for C, and vice versa. The payoff matrix is set up from R's point of view; thus, positive entries are gains for R (and losses for C), whereas negative entries are losses for R (and gains for C).

The optimal strategy for R is to play the row whose smallest value is maximal among the smallest values of rows. C should play the column whose largest value is minimal among largest values of columns. If the intersection of R's row and C's column is the maximum of R's smallest values and the minimum of C's largest values, then this value is called a saddle point; it is also the value of the game. A game with a saddle point is strictly determined.

In games without saddle points, optimal strategies are mixed. Let $r$ be a row matrix of probabilities with which player R plays each row, and let $c$ be a column matrix of probabilities with which C plays the columns. Then the expected value of the game corresponding to the mixed strategies r and c is rPc, where P is the payoff matrix.

To find the optimal mixed strategies for R and C, add a value to each entry of the payoff matrix, $P$ to obtain $P'$. Solve the linear programming problem:

maximize $M = x_1 + x_2 + x_3 + ... + x_n = [1\ 1\ ...\ 1]X$

subject to $P'X \geq \begin{bmatrix} 1 \\ 1 \\ 1 \\ . \\ . \\ . \\ 1 \end{bmatrix}$, $X = \begin{bmatrix} x_1 \\ x_2 \\ . \\ . \\ . \\ x_n \end{bmatrix} \geq 0$. Let $Y = [y_1 y_2 ... y_m]$ be the solution of the dual problem.

Then $r = \dfrac{Y}{y_1 + y_2 + ... + y_m}$, and $c = \dfrac{X}{x_1 + x_2 + ... + x_n}$ are the optimal strategies.

# Test Yourself

1. Decide whether each of the following is the matrix for a strictly determined game. If so, find the value of the game; if not, explain why not.

   a. $\begin{bmatrix} 1 & -1 \\ 4 & 2 \end{bmatrix}$

   b. $\begin{bmatrix} 1 & 3 & 5 \\ -2 & 2 & -3 \\ 4 & 0 & -1 \end{bmatrix}$

   c. $\begin{bmatrix} 1 & 0 & 1 \\ -1 & 2 & 4 \end{bmatrix}$

   d. $\begin{bmatrix} 2 & -1 \\ 3 & 0 \\ 2 & -2 \end{bmatrix}$

2. R and C play the game Scissors, Paper, Stone. Scissors wins $1 over paper, stone wins $2 over paper, and stone wins $3 over scissors.

   In case of a tie, there is no loss.

   a. Set up the payoff matrix.

   b. Decide whether the game is strictly determined. Explain.

   c. If the game is strictly determined, find the value of the game.

   d. Find the expected value of the game if R plays the mixed strategy r = [.5 .25 .25] and C plays the strategy c = $\begin{bmatrix} .2 \\ .5 \\ .3 \end{bmatrix}$.

3. R has cards with 2, 3, and 4. C has cards with 3, 4, and 5. Each plays one card. If the sum is odd, R receives the amount of the sum from C. If the sum is even, C receives the amount of the sum from R.

   a. Find the payoff matrix.

   b. Is the game strictly determined? Explain.

   c. If the game is strictly determined, find the value of the game.

   d. Find the expected value of the game if R plays the mixed strategy r = [.5 .25 .25] and C plays the strategy c = $\begin{bmatrix} .2 \\ .5 \\ .3 \end{bmatrix}$.

4. A game has payoff matrix $\begin{bmatrix} 2 & 0 \\ -2 & 1 \end{bmatrix}$.

   a. If R knows C is playing c = $\begin{bmatrix} .3 \\ .7 \end{bmatrix}$, would it be better for R to play [.5 .5] or [.8 .2]?

   b. If C knows R is playing r = [.4 .6] , would it be better for C to play $\begin{bmatrix} .2 \\ .8 \end{bmatrix}$ or $\begin{bmatrix} .5 \\ .5 \end{bmatrix}$ ?

   c. Use the geometric method to find optimal strategies for R and C.

5. A game has payoff matrix $\begin{bmatrix} 4 & 2 & -8 \\ 2 & -3 & 1 \end{bmatrix}$.

   a. If R knows C is playing c = $\begin{bmatrix} .4 \\ .3 \\ .3 \end{bmatrix}$, would R be better off playing [.5 .5] or [.1 .9]?

   b. If C knows R is playing r = [.4 .6], would C be better off playing c = $\begin{bmatrix} .1 \\ .5 \\ .4 \end{bmatrix}$ or c = $\begin{bmatrix} .2 \\ .2 \\ .6 \end{bmatrix}$ ?

   c. Use the simplex method to find optimal strategies for R and C.

# ✔ Check Yourself

1.  a. $\begin{bmatrix} 1 & -1 \\ 4 & 2 \end{bmatrix}$ First, R finds the smallest number in each row $\begin{bmatrix} -1 \\ 2 \end{bmatrix}$ and plays the row with the larger number:

the 2 in row 2. Then C finds the largest number in each column [4 2], and plays the column with the smaller of these: the 2 in column 2. The 2 in position (2, 2) is the choice of both R and C, so the game is strictly determined and has value 2.

   b. $\begin{bmatrix} 1 & 3 & 5 \\ -2 & 2 & -3 \\ 4 & 0 & -1 \end{bmatrix}$ R finds the smallest number in each row $\begin{bmatrix} 1 \\ -3 \\ -1 \end{bmatrix}$ and selects the row with the largest of these:

the 1 in row 1. C finds the largest number in each column [4 3 5], and selects the column with the smallest of these: the 3 in column 2. However, R and C do not choose the same entry, so this game is not strictly determined.

   c. $\begin{bmatrix} 1 & 0 & 1 \\ -1 & 2 & 4 \end{bmatrix}$ R finds the smallest value in each row $\begin{bmatrix} 0 \\ -1 \end{bmatrix}$ and selects the larger one: the 0 in row 1. C finds

the smallest value in each column: [−1 0 1], and selects the largest of these: the 1 in column 3. R and C do not select the same entry, so this game in not strictly determined.

   d. $\begin{bmatrix} 2 & -1 \\ 3 & 0 \\ 2 & -2 \end{bmatrix}$ R finds the smallest number in each row $\begin{bmatrix} -1 \\ 0 \\ -2 \end{bmatrix}$ and selects the largest of these: the 0 in row 2.

C finds the largest number in each column [3 0], and selects the smaller of these: the 0 in column 2. R and C select the same entry, the 0 in the (2, 2) position, so this game is strictly determined.
**(Strictly determined games)**

2.  
$$\begin{array}{c} \\ \text{Sc} \\ \text{a. P} \\ \text{St} \end{array} \begin{array}{ccc} \text{Sc} & \text{P} & \text{St} \\ \begin{bmatrix} 0 & 1 & -3 \\ -1 & 0 & -2 \\ 3 & 2 & 0 \end{bmatrix} \end{array}$$

   b. R finds the smallest number in each row $\begin{bmatrix} -3 \\ -2 \\ 0 \end{bmatrix}$ and selects the largest of these: the 0 in the third row. C

finds the largest number in each column [3 2 0], and selects the smallest of these: the 0 in column 3. Thus, R and C select the 0 in the (3, 3) position, and the game is strictly determined.

   c. The value of the game is 0. **(Strictly determined games)**

   d. $E = [.5\ .25\ .25] \begin{bmatrix} 0 & 1 & -3 \\ -1 & 0 & -2 \\ 3 & 2 & 0 \end{bmatrix} \begin{bmatrix} .2 \\ .5 \\ .3 \end{bmatrix} = [.5\ .25\ .25] \begin{bmatrix} -.4 \\ -.8 \\ 1.6 \end{bmatrix} = 0.$ **(Mixed strategies)**

3.  
$$\begin{array}{c} \\ 2 \\ \text{a. } 3 \\ 4 \end{array} \begin{array}{ccc} 3 & 4 & 5 \\ \begin{bmatrix} 5 & -6 & 7 \\ -6 & 7 & -8 \\ 7 & -8 & 9 \end{bmatrix} \end{array}$$

b. R finds the smallest number in each row $\begin{bmatrix} -6 \\ -8 \\ -8 \end{bmatrix}$ and selects the largest of these: the $-6$ in row 1. C finds the largest number in each column [7 7 9], and selects the smallest of these. C has a choice of the 7 in column 1 or the 7 in column 2. In neither case do R and C select the same entry, so this game is not strictly determined.

c. The game is not strictly determined. **(Strictly determined games)**

d. $E = [.5\ .25\ .25] \begin{bmatrix} 5 & -6 & 7 \\ -6 & 7 & -8 \\ 7 & -8 & 9 \end{bmatrix} \begin{bmatrix} .2 \\ .5 \\ .3 \end{bmatrix} = [.5\ .25\ .25] \begin{bmatrix} .1 \\ -.1 \\ .1 \end{bmatrix} = .05$. **(Mixed strategies)**

4.  a. If R plays [.5 .5], the expected value of the game is $[.5\ .5] \begin{bmatrix} 2 & 0 \\ -2 & 1 \end{bmatrix} \begin{bmatrix} .3 \\ .7 \end{bmatrix} = [.5\ .5] \begin{bmatrix} .6 \\ .1 \end{bmatrix} = .35$, whereas if R plays [.8 .2], the expected value is $[.8\ .2] \begin{bmatrix} 2 & 0 \\ -2 & 1 \end{bmatrix} \begin{bmatrix} .3 \\ .7 \end{bmatrix} = [.8\ .2] \begin{bmatrix} .6 \\ .1 \end{bmatrix} = .5$. Since R seeks the larger values, R should play r = [.8 .2] .

b. If C plays $\begin{bmatrix} .2 \\ .8 \end{bmatrix}$, the expected value of the game is $[.4\ .6] \begin{bmatrix} 2 & 0 \\ -2 & 1 \end{bmatrix} \begin{bmatrix} .2 \\ .8 \end{bmatrix} = [-.4\ .6] \begin{bmatrix} .2 \\ .8 \end{bmatrix} = .4$, and if C plays $\begin{bmatrix} .5 \\ .5 \end{bmatrix}$, the expected value is $[.4\ .6] \begin{bmatrix} 2 & 0 \\ -2 & 1 \end{bmatrix} \begin{bmatrix} .5 \\ .5 \end{bmatrix} = [-.4\ .6] \begin{bmatrix} .5 \\ .5 \end{bmatrix} = .1$. Since C seeks to minimize the expected value, C should play $\begin{bmatrix} .5 \\ .5 \end{bmatrix}$. **(Mixed strategies)**

c. First, add 3 to each entry of the payoff matrix so that all entries will be positive, obtaining $P' = \begin{bmatrix} 5 & 3 \\ 1 & 4 \end{bmatrix}$. To find the optimal strategy for C, maximize M = x + y subject to

$$5x + 3y \leq 1$$
$$x + 4y \leq 1$$
$$x \geq 0, y \geq 0$$

The feasible region has corners (0, 0), (0, 1/4), (1/5, 0) and (1/17, 4/17). The maximum is M = 5/17, so that $c_1 = \dfrac{1/17}{5/17} = 1/5 = .2$, and $c_2 = \dfrac{4/17}{5/17} = 4/5 = .8$. So that $c = \begin{bmatrix} .2 \\ .8 \end{bmatrix}$.

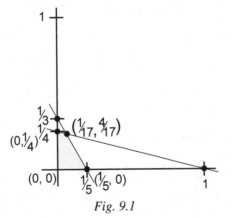

*Fig. 9.1*

To find the optimal strategy for R, minimize m = x + y subject to

$5x + y \geq 1$

$3x + 4y \geq 1$

$x \geq 0, \ y \geq 0$

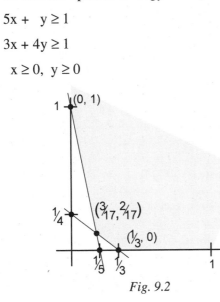

*Fig. 9.2*

The feasible region has corners (0, 1), (1/3, 0), and (3/17, 2/17), and the minimum value is m = 5/17.

Thus, $r = \dfrac{3/17}{5/17} = 3/5 = .6$ and $r_2 = \dfrac{2/17}{5/17} = 2/5 = .4$, so r = [.6 .4]. The expected value of the game is

$[.6 \ .4]\begin{bmatrix} 2 & 0 \\ -2 & 1 \end{bmatrix}\begin{bmatrix} .2 \\ .8 \end{bmatrix} = .4.$ **(Optimal strategies)**

5.   a. If R plays [.5 .5], the expected value of the game is $[.5 \ .5]\begin{bmatrix} 4 & 2 & 8 \\ 2 & -3 & 1 \end{bmatrix}\begin{bmatrix} .4 \\ .3 \\ .3 \end{bmatrix} = [.5 \ .5]\begin{bmatrix} -.2 \\ .2 \end{bmatrix} = 0.$ If R plays

[.1 .9], then the expected value of the game is $[.1 \ .9]\begin{bmatrix} 4 & 2 & -8 \\ 2 & -3 & 1 \end{bmatrix}\begin{bmatrix} .4 \\ .3 \\ .3 \end{bmatrix} = [.1 \ .9]\begin{bmatrix} -.2 \\ .2 \end{bmatrix} = .16.$ R seeks to

maximize the value, so R should play [.1 .9].

b. If C plays $\begin{bmatrix} .1 \\ .5 \\ .4 \end{bmatrix}$, the expected value of the game is $[.4 \ .6]\begin{bmatrix} 4 & 2 & -8 \\ 2 & -3 & 1 \end{bmatrix}\begin{bmatrix} .1 \\ .5 \\ .4 \end{bmatrix} = [2.8 \ -1 \ -2.6]\begin{bmatrix} .1 \\ .5 \\ .4 \end{bmatrix} = -1.26.$

If C plays $\begin{bmatrix} .2 \\ .2 \\ .6 \end{bmatrix}$, the expected value is $[.4 \ .6]\begin{bmatrix} 4 & 2 & -8 \\ 2 & -3 & 1 \end{bmatrix}\begin{bmatrix} .2 \\ .2 \\ .6 \end{bmatrix} = [2.8 \ -1 \ -2.6]\begin{bmatrix} .2 \\ .2 \\ .6 \end{bmatrix} = -1.2.$

Since C seeks to minimize, C should play $\begin{bmatrix} .1 \\ .5 \\ .4 \end{bmatrix}$. **(Mixed strategies)**

c. First, add 8 to each of the entries of P so that there are no negative values. $P' = \begin{bmatrix} 12 & 10 & 0 \\ 10 & 5 & 9 \end{bmatrix}$.

Solve maximize $M = x + y + z$

subject to $\quad 12x + 10y \quad\quad \leq 1$

$\quad\quad\quad\quad 10x + 5y + 9z \leq 1$

$\quad\quad\quad\quad x \geq 0,\ y \geq 0,\ z \geq 0$

$$\begin{bmatrix} x & y & z & u & v & M & | & \text{sol} \\ 12 & 10 & 0 & 1 & 0 & 0 & | & 1 \\ 10 & 5 & 9 & 0 & 1 & 0 & | & 1 \\ - & - & - & - & - & - & | & - \\ -1 & -1 & -1 & 0 & 0 & 1 & | & 0 \end{bmatrix} \rightarrow \begin{bmatrix} x & y & z & u & v & M & | & \text{sol} \\ 12 & 10 & 0 & 1 & 0 & 0 & | & 1 \\ {}^{10}\!/_9 & {}^5\!/_9 & 1 & 0 & {}^1\!/_9 & 0 & | & {}^1\!/_9 \\ - & - & - & - & - & - & | & - \\ {}^1\!/_9 & -{}^4\!/_9 & 0 & 0 & {}^1\!/_9 & 1 & | & {}^1\!/_9 \end{bmatrix} \rightarrow \begin{bmatrix} x & y & z & u & v & M & | & \text{sol} \\ {}^6\!/_5 & 1 & 0 & {}^1\!/_{10} & 0 & 0 & | & {}^1\!/_{10} \\ {}^4\!/_9 & 0 & 1 & -{}^1\!/_{18} & {}^1\!/_9 & 0 & | & {}^1\!/_{18} \\ - & - & - & - & - & - & | & - \\ {}^{29}\!/_{45} & 0 & 0 & {}^2\!/_{45} & {}^1\!/_9 & 1 & | & {}^7\!/_{45} \end{bmatrix}$$

Therefore, $x = 0$, $y = 1/10$, $z = 1/18$, $u = 2/45$, and $v = 1/9$. Therefore, $c_1 = 0$, $c_2 = \dfrac{{}^1\!/_{10}}{{}^7\!/_{45}} = 9/14$ and $c_3 = \dfrac{{}^1\!/_{18}}{{}^7\!/_{45}} =$

$5/14$, and C's optimal strategy is $c = \begin{bmatrix} 0 \\ {}^9\!/_{14} \\ {}^5\!/_{14} \end{bmatrix}$. Also, $r_1 = \dfrac{{}^2\!/_{45}}{{}^7\!/_{45}} = 2/7$ and $r_2 = \dfrac{{}^1\!/_9}{{}^7\!/_{45}} = 5/7$, and R's optimal

strategy is $r = [2/7\ 5/7]$. The value of the game is $[2/7\ 5/7] \begin{bmatrix} 4 & 2 & -8 \\ 2 & -3 & 1 \end{bmatrix} \begin{bmatrix} 0 \\ {}^9\!/_{14} \\ {}^5\!/_{14} \end{bmatrix} = [18/7\ -11/7\ -11/7] \begin{bmatrix} 0 \\ {}^9\!/_{14} \\ {}^5\!/_{14} \end{bmatrix} =$

$-154/98 = -11/7$. **(Optimal strategies)**

# Grade Yourself

Circle the numbers of the questions you missed, then fill in the total incorrect for each topic. If you answered more than three questions incorrectly, you need to focus on that topic. (If a topic has less than three questions and you had at least one wrong, we suggest you study that topic also. Read your textbook or a review book, or ask your teacher for help.)

## Subject: Game Theory

| Topic | Question Numbers | Number Incorrect |
|---|---|---|
| Strictly determined games | 1, 2a-c, 3a-c | |
| Mixed strategies | 2d, 3d, 4a-b, 5a-b | |
| Optimal strategies | 4c, 5c | |

# Also Available

*. . . and many others to come!*